D0816427

ENCYCLOPEDIA OF
AUSTRALIAN
ANIMALS

MAMMALS

ENCYCLOPEDIA OF AUSTRALIAN ANIMALS

MAMMALS

RONALD STRAHAN

SERIES EDITOR RONALD STRAHAN

THE NATIONAL PHOTOGRAPHIC INDEX OF AUSTRALIAN WILDLIFE
THE AUSTRALIAN MUSEUM

Angus&Robertson
An imprint of HarperCollins*Publishers*

Photographs appearing on the preliminary pages:
x Queensland Tube-nosed Bat (Nyctimene robinsoni) *R & A Williams*
xiii Spotted Cuscus (Phalanger maculatus) *G Schick*
xv Quokka (Setonix brachyurus) *J Lochman*
xvi-1 Numbat (Myrmecobius fasciatus) *R Whitford*

AN ANGUS & ROBERTSON BOOK
An imprint of HarperCollinsPublishers

First published in Australia in 1992 by
CollinsAngus&Robertson Publishers Pty Limited (ACN 009 913 517)
A division of HarperCollinsPublishers (Australia) Pty Limited
25-31 Ryde Road, Pymble NSW 2073, Australia

HarperCollinsPublishers (New Zealand) Limited
31 View Road, Glenfield, Auckland 10, New Zealand

HarperCollinsPublishers Limited
77-85 Fulham Palace Road, London W6 8JB, United Kingdom

© *The Australian Museum Trust 1992*

This book is copyright.
Apart from any fair dealing for the purposes of private study,
research, criticism or review, as permitted under the Copyright
Act, no part may be reproduced by any process without written
permission. Inquiries should be addressed to the publishers.

National Library of Australia
Cataloguing-in-Publication data:

Encyclopedia of Australian animals.
  Includes index.
  ISBN 0 207 17378 8 (Mammals).
  1. Zoology—Australia—Encyclopedias.
  2. Animals—Encyclopedias. I. Strahan,
  Ronald, 1922—

Layout by Sue Edmonds
Typeset by Midland Typesetters
Printed in Hong Kong
5 4 3 2 1
96 95 94 93 92

FOREWORD

The National Photographic Index of Australian Wildlife is a special project of the Australian Museum. Its initial purpose was to collect and preserve outstanding colour photographs of living animals, and thereby to supplement the Museum's collections of preserved specimens, but it soon became apparent that the photographs could be the basis of outstanding books. In turn, these books would preserve the photographic images, make them available to a wide public, and return revenue to the operation.

The Index served first as the core of the *Reader's Digest Complete Book of Australian Birds*, which has passed through numerous reprintings and appeared as a new edition in 1990. From its first appearance, it has been the most comprehensive and authoritative of popular references to the birds of Australia.

This was followed in 1983 by the *Australian Museum's Complete Book of Australian Mammals*, published by Angus & Robertson. Filling a vacuum in both the professional and amateur literature of these animals, it is an even more significant work.

Meanwhile, steadily working away, the Index continues its association with Collins/Angus & Robertson in the production of ten volumes on the birds of Australia, seven of which have so far been published. Additionally, the collections of the Index, representing the work of the best of Australia's wildlife photographers, have been used by hundreds of authors and publishers to illustrate other works.

The four volumes of the *Encyclopedia of Australian Animals* document all Australian species of frogs, reptiles, birds and mammals and provide an invaluable reference for the increasing number of people in Australia, and around the world, who are concerned with our unique fauna.

The contributions of the Index represent but one aspect of the mission of the Australian Museum, which is to increase understanding of our natural environment and cultural heritage. We aim to be a catalyst in developing and changing people's attitudes through exhibitions, education programmes, lectures and publications. The *Encyclopedia of Australian Animals* is a valuable contribution to this mission and I am delighted by this latest product of the relationship between the Index and its publisher.

Des Griffin AM
Director
The Australian Museum

SPONSORS

The following have contributed to the cost of production of this work.

E.M. & M.J. Abrahams
J. & J. Allison
Arnott's Biscuits Ltd
A.W. Auldist
C.N. Banks
G.W.E. Barraclough
H. Barry
P.J. & J.Y. Bath
K.A. Blofeld
Dr. M. Bonnin
C. & T. Britton
G. Broinowski
J. & J. Broinowski
S. Broinowski
H.G. Brooks
P.J. Buckley
J., N., S., P., H. & H. Burton Taylor
A. Byrne
Sir Bede Callaghan
D.M. Carment
J. Champ
A. Charles
Clampett Properties Ltd
M.D. Cobcroft
B. & J. Coghlan
L.R. & P. Comben
K.H. Cousins
S. & M. Crouch
Mr and Mrs Cumbrae Stewart
Dalgety Farmers Ltd
P. & I. Davidson
S.G. Eagles
P. Edwards, E. & T. Frankland
J.O. Fairfax
Sir Vincent & Lady Fairfax
B.A. Farrell
C.A. Fay
Dr. M.A. Feilman OBE
R.A. Field

T. Florin
D. Ford
B. & P. France
K., R. & R. Frankland
K.I. Frecker
M.D. Frecker
E.W. Gibson
Sir Archibald Glenn
B. Goldrick
M.R. & B. Gordon
J.D. Gorter Pty Ltd
Gould League of NSW
Gould League of W.A.
I.G. Graham
R.W. Greaves
Sir David & Lady Griffin
R.M. Griffin
N. & E. Haines
C.M. Hall
J.D. Hamman
M. Hamilton
K.C. Hammer
R. & P. Harrison
H.R. Hawkeswood
Dr. J. & M. Hazel
S.M. Hicks
A. Holmes
R.M. Howarth
S. Hughes
Hunter Wetland Trust
T.D.E. Hyde
D. Inglis
M.M. Johnson
E. & T. Karplus
A.C. Keating
G.M. King
H.W. & M.A. Kinnersley
L.B. Kirk
L. Kramer

Y. Lewis
B. & N. Lithgow
A.F. Little
K. Lyall
K.A. Macpherson
H. & F. MacLachlan
A. McBride
F.L. Mahoney
M.J. Mashford
S. & R. Miles
P. Miles
L. & J. Minnett
B.N. Morrison
J. & M. Munro
Dr. M. Norst
Pancontinental Mining Ltd
L. Papi
Peddle, Thorpe & Walker P/L
Progressive Mortgage Co. Ltd.
Sir John & Lady Proud
Renison Goldfields Consolidated Ltd
N. Robinson
Royal Zoological Society of NSW
RSPCA (NSW)
C. & M. Ryman
P. & J. Sayers
R. & R. Schmidt
B.M. Scott
M. & R. Shepherd
R.A. Simpson
D. Solomon and family
South Australian Museum
Dr. S. Stevens
B. Stevenson
Sir Edward Stewart
Sunshine Foundation
Sir James Vernon
J.S. & M. Wilkey
R. & T. Yates

CONTENTS

For individual species listing, see
the Index of Common Names or
the Index of Scientific Names.

PREFACE

The National Photographic Index of Australian Wildlife was established by the Australian Museum in 1969, originally to create a set of reference photographs of all the Australian birds. These are being used in the production of a series of richly illustrated volumes, seven of which have so far been published by Collins/Angus & Robertson.

Expansion of the collection to include mammals provided a wealth of photographs to illustrate *The Australian Museum Complete Book of Australian Mammals*, the text of which was written by 110 experts. It is currently the standard work on Australian mammal species, accessible to the layman, but also referred to consistently by professional zoologists: no other source is more frequently cited in the current literature of Australian mammalogy.

In recent years, the Index has extended its scope to include the Australian frogs and reptiles. Thanks to the cooperation of hundreds of amateur and professional herpetologists, its collection of colour photographs of these animals is now the most comprehensive in existence. It therefore seemed appropriate to produce a series that would combine information and illustrations of all of these animals.

Zoologists tend to write about their areas of expertise but most members of the public have general interests. To meet this need, we have produced four volumes in which, for the first time, all of the basically four-legged Australian animals (tetrapods) have been treated in much the same way.

One need thumb through only a few pages to recognise that our knowledge of many species is remarkably slight. In some instances, we know almost nothing of the way of life of an animal, it being represented by a few museum specimens. It has been disappointing to the authors that we have been unable to provide an interesting account and colour photograph of every species, but the gaps may serve a useful purpose in drawing attention to the need for further research on the Australian fauna.

In a work of this nature, it is impossible to give credit for the sources of every item of information compressed into its pages: these represent the cumulative work of thousands of amateur and professional zoologists over the course of the past two centuries. However, since the work would not have been contemplated in the absence of the Index, it is pertinent to thank the hundreds of photographers who have contributed to the Index collections and Donald Trounson, who founded the Index and made its enterprises possible.

The task of assembling the photographs for the four volumes was begun by the Archivist, Heather Lawrence, and carried through by the Collections Manager, Sally Bird. Joy Coghlan made sense of successively edited drafts of the text.

Finally, I acknowledge the many sponsors who contributed to the great cost of producing the work.

Ronald Strahan
Editor-in-Chief
National Photographic Index of Australian Wildlife

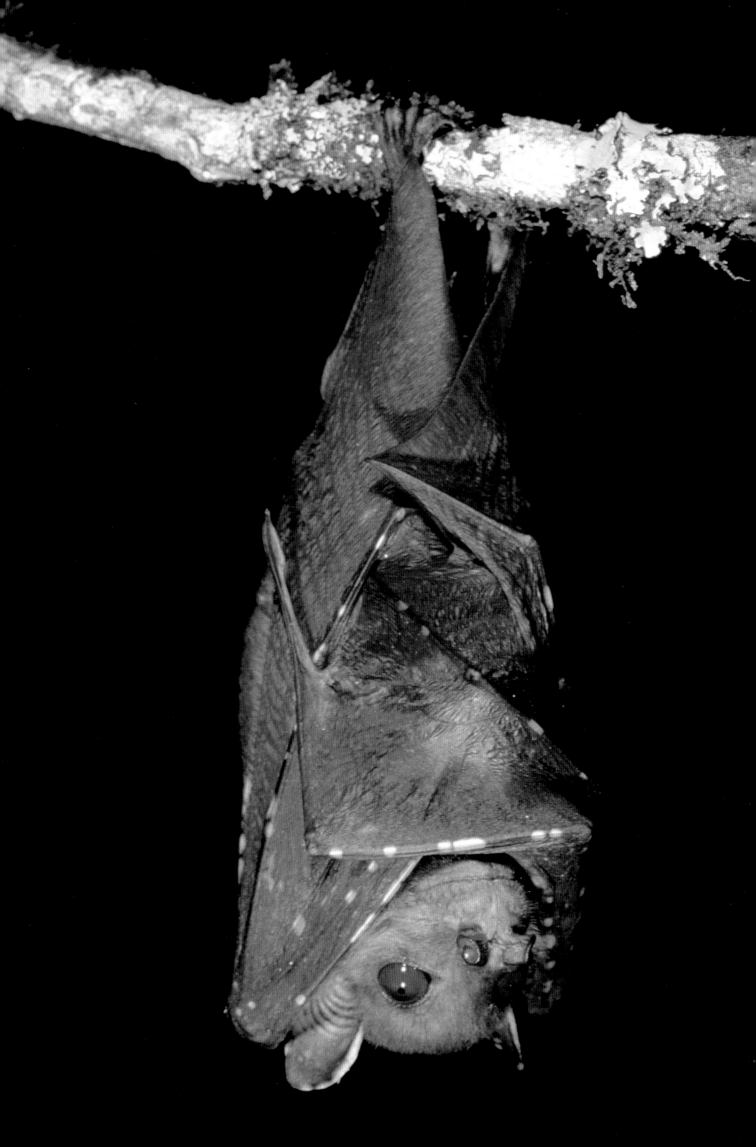

INTRODUCTION

This book is one of four volumes that comprise an encyclopedia of those Australian animals that have a backbone and four legs: the encyclopedia also includes all those animals such as snakes, birds and bats that have evolved from four-legged ancestors. These volumes, and their contents, are arranged according to their zoological classification.

CLASSIFICATION

Zoological classification is based on a system of narrower and narrower categories, similar to those in the address of an overseas letter. Postal workers first sort letters in terms of continent and country, then cities or towns, street and number, and finally the surname and personal name of the addressee. This may be compared, for example, with the classification of the first species in this volume, the rat-sized Mulgara of the central deserts. Having a backbone, it is a member of the Subphylum Vertebrata (not a worm, mollusc, arthropod, etc.). Its coat of hair defines it as belonging to the Class Mammalia (not an amphibian, reptile or bird). The small size and undeveloped state of the newborn define it as a member of the Subclass Marsupialia (not a monotreme or a eutherian). Its Australian distribution and the structure of its sperms indicate it to be a member of the Cohort Australidelphia (not a South American marsupial). Carnivorous dentition and other anatomical characters confirm it is a member of the Order Dasyuromorphia (not a bandicoot, possum or kangaroo) and other details confine it to the Family Dasuryidae (not a Thylacine or Numbat). Skull characters, teeth and tail define it as the genus *Dasycercus*, which has only one species, *cristicauda*.

Classification is a means of sorting and labelling animals but it has another important function in indicating evolutionary relationships. When a zoologist places two or more species in the same genus, this is the expression of an opinion that these are closely related to each other and share a common ancestor that is different from that of the species in other genera. Relationship is also implied when genera are placed in the same family, or families in the same order. If we had complete knowledge of all living and extinct species, we might be able to construct a classification that reflected the entire history of animal evolution, but lack of knowledge forces zoologists to make decisions on incomplete information. Consequently, schemes of classification are always open to revision in response to new information or opinions.

ZOOLOGICAL NAMES

Fundamental to the science of zoology is the binominal (two-name) system, according to which every species has a name composed of two parts, the genus name and the specific name. A genus name is a noun and can be used by itself, but a species name functions as an adjective and cannot stand alone.

A widespread species often includes populations that differ in size, proportions or coloration at the extremes of its range (north to south, east to west, wet to dry, or low to high altitude). Where the transition is continuous and gradual from one form to another, it is referred to as a cline. Where the variation is discontinuous and animals from one area differ recognisably from those in another, we refer to each form as a **subspecies**. Where one subspecies adjoins another, there are usually intermediate forms. Subspecies are named by adding a third (subspecific) name to that of the species.

PRONUNCIATION

Because most zoological names are constructed from Latin or Greek roots, they should be pronounced and stressed in such a way as to retain the identity of these elements. There are no absolute rules, but the pronunciation recommended in this encyclopedia is reasonably consistent and international. Consonants are given their usual English values: a limited set of vowels is pronounced as below.

ay as in bay	*ie* as in die	*aw* as in law
a as in bat	*i* as in bit	*ue* as in sue
ah as in bah	*air* as in fair	*u* as in but
ee as in bee	*oh* as in doh	*oo* as in good
e as in bet	*o* as in dot	*ow* as in cow
er as in fern	*or* as in for	*oy* as in boy

A point to be noted is that proper names retain their identity, even when Latinised. Thus *burrelli* (after a Mr Burrell) is bu'-rel-ee, not bu-rel'-ee; *sladei* (after a Mr Slade) is slay'-dee; *godmani* (after a Mr Godman) is god'-mun-ee, not god-mah'-nee.

COMMON NAMES

Most Australian birds and mammals have agreed common names but only the more common frogs and reptiles are so blessed. The authors of this work have attempted to provide a consistent set of common names for every Australian tetrapod, choosing among alternatives when these are available and creating names where none previously existed.

PROSPECTS OF SURVIVAL

Human activities have led to the extinction of many vertebrate species and an even greater number are endangered or vulnerable. Nevertheless, many species are doing well at present. To deal with this range of situations, this encyclopedia introduces a three-fold system of categorisation of survival status comprising estimates of range and of abundance with a prognosis for each species.

Distribution

It is a reasonable assumption that a widely distributed species is more secure than one limited to a small area. Distribution is scored in six categories based on distribution maps and, since such maps always include areas where the species does not occur, these figures are always overestimates.

The principle underlying the scoring system for distribution is that the smaller the area occupied by a species, the greater the significance of that area. The scale therefore expands exponentially, beginning with small increases and ending with very large ones.

SCORING OF DISTRIBUTION

Less than 10,000 square kilometres	10,000–30,000 square kilometres
30,000–100,000 square kilometres	100,000–300,000 square kilometres
300,000–1,000,000 square kilometres	More than 1,000,000 square kilometres

---

Abundance

In principle, abundance should be a measure of the number of individuals in a given area (and abundance multiplied by distribution would represent the population of a species). We seldom have the resources to conduct such counts but experienced observers can reach reasonable agreement on the six categories of abundance set out below.

Our best information on abundance comes from zoologists working in such institutions as natural history museums, national parks and wildlife services, the CSIRO, and universities. Such people are familiar with a range of species and are frequently in the bush. Amateur ornithologists and

herpetologists are also valuable sources of information. The scoring system depends upon the opinions of experienced observers who are familiar with the species in question and judgment that a species is abundant does not necessarily mean that it will appear to be so to an untrained person. In the definition that follows, "locatable" means seen, heard, or judged to be present by nests, burrows, scrapings, dung, food scraps etc. "Appropriate times" may mean the time of day when the species is active or the time of year when its presence becomes obvious.

SCORING OF ABUNDANCE

VERY RARE	So infrequently located or trapped, despite considerable efforts, that the possibility of extinction cannot be excluded.
RARE	Known to be present within the distribution area but seldom located or trapped, despite considerable effort. May not be located over periods of several years.
VERY SPARSE	As for "sparse" but infrequently located, usually after considerable effort; apparently absent from many appropriate habitats.
SPARSE	Known to be present and frequently located at appropriate times in some appropriate habitats, but usually after some effort; if trappable, frequently absent from trap-lines set at such times and places.
COMMON	Usually locatable at appropriate times in most appropriate habitats; if trappable, usually represented to some extent in most trap-lines set at such times and places.
ABUNDANT	Locatable at all appropriate times in all appropriate habitats; if trappable, well represented in any trap-line set at such times and places.

Survival Status

Because conservation agencies have been mainly concerned with species at risk, evaluation of survival status has usually been restricted to these categories. The novelty in the approach taken here is recognition of two categories—"secure" and "possibly secure"—representing antitheses to "possibly extinct" and "presumed extinct". Before allotting a species to one of these categories, we have considered its past and present distribution and abundance, and those environmental and human factors that are currently acting to its detriment or benefit. Prognoses of survival status are scored as follows.

SCORING OF SURVIVAL STATUS

PRESUMED EXTINCT	No confirmation of the existence of the species in the wild for 50 years or more.
POSSIBLY EXTINCT	So infrequently located that the existence of the species in the wild has not been confirmed for up to 50 years.
ENDANGERED	Declining in distribution and/or abundance to such an extent that, without positive action to halt or reverse the trend, the species appears likely to become extinct in the near future.
POSSIBLY ENDANGERED	Declining in distribution and/or abundance but population still large and viable.
VULNERABLE	Currently of satisfactory distribution and/or abundance but foreseeable pressures could put the species at risk.
PROBABLY SECURE	Probably as below but insufficient data to be certain.
SECURE	No existing or foreseeable threat to continuance of the species.

Because this is the first attempt to survey the survival status of the Australian tetrapods, these evaluations will inevitably include some errors of fact or judgment. They should therefore be used as first approximations and as the basis for discussion.

In an overall sense, they must also be regarded as optimistic and good for only a few decades. A pessimistic approach, postulating unrestricted expansion of human populations, critical levels of environmental pollution, a severe "greenhouse" effect and possibly a "nuclear winter", would require the classification of virtually every tetrapod, including humans, as vulnerable or endangered.

CLASS

Mammalia

(mam-ah'-lee-ah: "mammals")

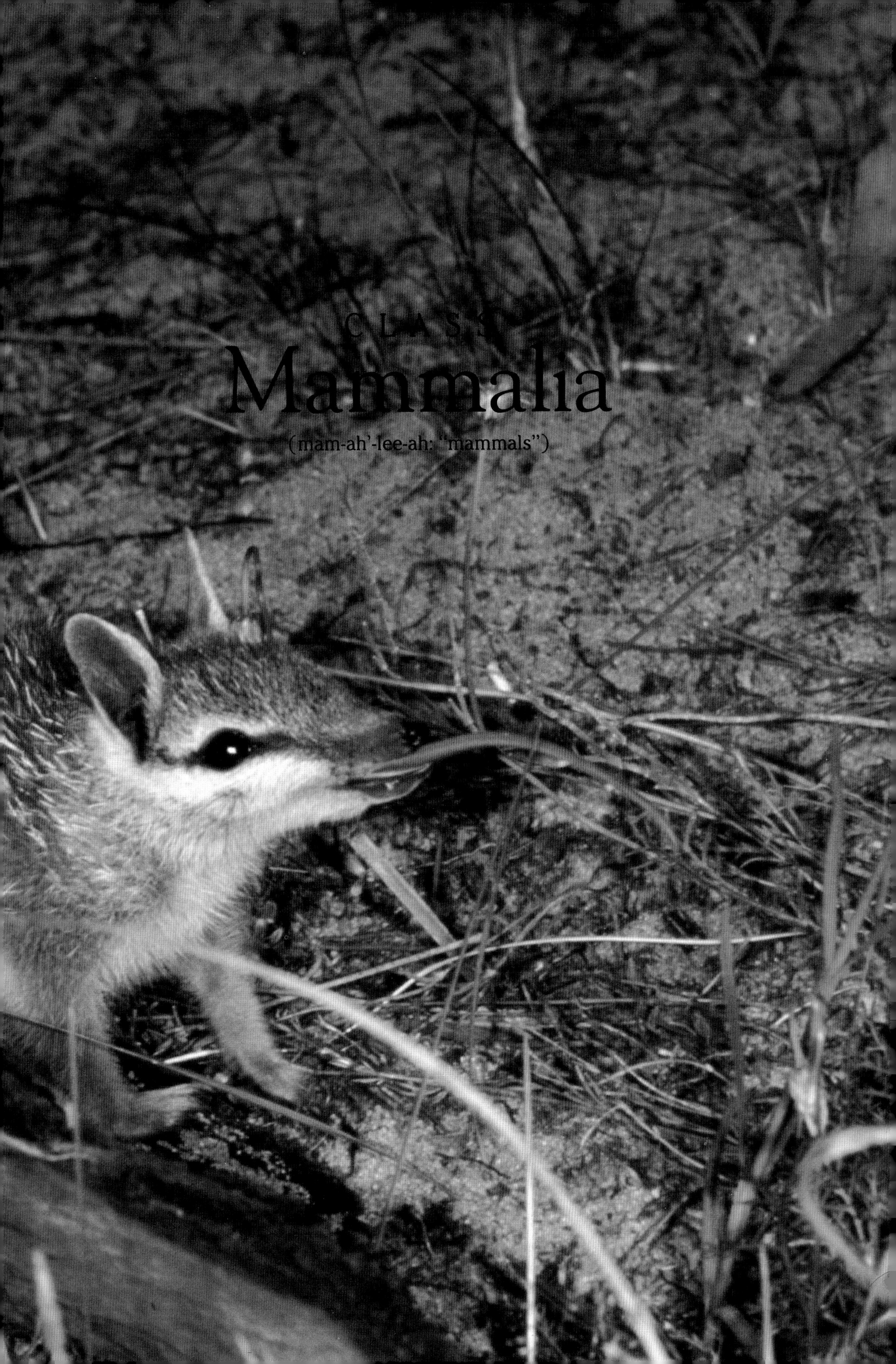

Although the evolutionary transition from reptiles to mammals was a slow and gradual process, all of the intermediate forms are long extinct and modern mammals are readily distinguished from the surviving reptiles by many clear differences. The most obvious of these is that mammals have hair, usually as a full covering of fur but sometimes incomplete, as in humans, or reduced to a few bristles, as in whales. The main function of hair is to provide thermal insulation, retarding the loss of heat from the body, and this is related to the warm-bloodedness of mammals (feathers provide similar insulation in the warm-blooded birds). Many reptiles are able to maintain a high, and remarkably constant, body temperature *during the day* by shuttling between hot and cool areas but they cool down at night. Apart from birds, mammals are unique in the animal kingdom in having such a high rate of metabolism that they continually produce an excess of heat; a constant warm body temperature being then maintained by the controlled loss of such heat by sweating, panting, adjusting the fur or seeking cool surroundings. Some very small mammals which are unable to balance their heat-loss by input of food, "switch off" their temperature-control systems when resting and conserve energy by allowing the temperature of the body to fall almost to that of the surroundings: this is known as *torpor*. Others, including species as large as bears, go further and pass into a cool, quiescent state for periods of up to several months during the winter: this is known as *hibernation*.

Energy, whether to power the overall activities of a mammal or to produce heat, comes from what it eats. A mammal therefore has to consume much more food than a cold-blooded animal of equivalent size. In this sense, life is harder for a mammal than a reptile but there are corresponding advantages. The vast majority of mammals are active at night, when most reptiles have cooled down and become inactive. Moreover, mammals have spread into the coldest parts of the world, whereas reptiles are concentrated in the tropics and are rather sparsely represented in the cool temperate regions. A constant warm body temperature also provides the basis of a stable *internal environment* for the activities of the body, particularly the nervous system and muscles, which are able to function in a constant manner, irrespective of external conditions.

Warm-bloodedness did not evolve simply by an increase in the metabolic rate: many interlocking factors were involved in the origin of mammals from reptiles. For example, their greater need for food is associated with specialisation of the teeth for various functions. On each side of the upper and lower jaws of a typical mammal there are, from front to back, a variable number of simple, sharp incisors, used to hold on to prey or bite into plants; a strong, sharp canine, used to pierce prey or as a defensive/offensive weapon; several pre-molars, often used to shear flesh from the prey; and several molars that chop food into fragments. Small pieces of food are rapidly digested and release their energy much faster than the whole animals that reptiles usually swallow.

Mammals are vertebrates which *chew* their food: they have more powerful and elaborate jaw muscles than reptiles. These muscles require a strong, solid lower jaw and, in the evolution of the mammals, this strengthening involved a change in the hinge between the lower jaw and the cranium and the reduction of the lower jaw to a single bone on each side. Two small bones which had been involved in the original reptilian hinge became released from the jaw and incorporated into the middle ear as ossicles: whereas reptiles have only one ossicle, the *stapes*, mammals have three: the *stapes, incus* and *malleus*. Mammals still further increased the efficiency of their hearing by "inventing" the outer ear, or *pinna*, a funnel-like structure moved by muscles, which helps to locate the source of a sound.

In addition to a great input of food, warm-bloodedness requires a good supply of oxygen with which to oxidise ("burn up") that food. The lungs of mammals are powered partly by the ribs, as in reptiles, but more by a sheet of muscle, the *diaphragm*, which separates the chest from the abdomen. Yet another mammalian refinement is the division of the mouth cavity into two passageways by the *false palate*, permitting a mammal to continue breathing while it is chewing or swallowing: reptiles cannot do this.

Getting a lot of air into the lungs is not, in itself, sufficient to maintain a high temperature: oxygen has to pass into the blood and be distributed speedily and efficiently throughout the body. Mammalian blood can carry more oxygen than reptilian blood and the blood is oxygenated more efficiently because the heart is, in effect, two separate pumps; one directing blood from the body to the lungs; the other propelling it, at high pressure, from the lungs to the body. Among the living reptiles, only the crocodylians have a similar structure of the heart.

Most mammals give birth to live young but there is nothing intrinsically superior in this: many invertebrates, fishes, amphibians and reptiles do the same and one group of mammals, the monotremes, lays eggs. Rather more significant is the fact that members of the most successful group of mammals, the eutherians, have a very efficient placenta which provides such a close association of the embryonic and maternal blood systems that the mother is able to take responsibility for all of the nutritional, respiratory and excretory needs of the embryo in her womb. Thanks to the placenta, eutherian foetuses can grow to the size of a pony, calf, or whale pup by the time of their birth. Nevertheless, most newborn eutherian mammals are much smaller than a newly hatched emu, and newborn marsupials range in size from that of a grain of rice to a peanut kernel. Thus, contrary to a general belief, the success of mammals cannot be attributed simply, or even primarily, to their method of reproduction—even in respect of the largest mammals. Some egg-laying dinosaurs were far bigger than the largest mammals that ever dwelt on the land.

The name of the class Mammalia refers to the nourishment of young by milk produced from the mother's breasts (Latin *mamma*). Milk is a very convenient source of food and nursing creates a bond between mother and young—but most birds provide just as much care of their young as do mammals. The production of milk is certainly characteristic of mammals but it is not necessarily a biologically superior method of infant nutrition.

There are good reasons to suppose that, if the dinosaurs had not become extinct about 65 million years ago, the mammals might have remained a rather minor group of small, nocturnal, insectivorous tetrapods. The demise of the dominant reptiles (which included small as well as giant species) created an ecological "vacuum" and, in the competition to fill the gaps that had been created, the early mammals proved superior to the non-dinosaurian reptiles—possibly because of a combination of characteristics associated with warm-bloodedness.

There seems to have been little difference in the size of the brain of early mammals and their contemporary reptiles. However, as mammals evolved, their brains became bigger (and, presumably, better): the brains of modern mammals are significantly larger than those of the modern reptiles. Much of the present superiority of mammals can probably be attributed to their being "brighter" than reptiles, more able to adjust the behaviour of individuals in respect of individual experience, and somewhat less constrained by inherited instinct. (The warm-blooded birds did not take this path.)

Most of the major groups of mammals that have ever existed are now extinct. The surviving species fall into three major groups: the Prototheria, now represented only by the egg-laying Monotremata; the Marsupialia, which give birth to very small young with imperfectly developed hindlimbs; and the Eutheria, newborn young of which are proportionately larger than those of marsupials (usually very much so) and in which all four limbs are completely developed at the time of birth. It must be stressed that, while these differences provide convenient means of separating the three groups of modern mammals, they are probably not of fundamental significance in their relative degrees of success.

It is generally agreed that the Marsupialia and Eutheria are more closely related to each other than either is to the Prototheria: they are therefore often included within a superior category, the Theria. However, because the nomenclature of the higher categories of mammals is currently confused, the issue is avoided here.

Subclass PROTOTHERIA

(proh'-toh-thee'-ree-ah: "first mammals")

Most of the mammals that are classified as prototherians flourished between about 220 and 100 million years ago and are not represented in the Australian fossil record. Fossil monotremes are very sparse but include some large, recently extinct echidnas and a toothed platypus from about 15 million years ago. A monotreme jaw fragment from Australia is dated at about 100 million years and another from Argentina (made known in 1991) appears to be about 65 million years old. Contrary to long-held beliefs, monotremes are not restricted to the Australo-New Guinean Region.

Order MONOTREMATA

(mon'-oh-trem-ah'-tah: "one-holes")

Living monotremes comprise the Platypus and two species of echidnas. The name Monotremata refers to the single (cloacal) aperture in the female, through which faeces, urine and eggs are passed (as in reptiles, marsupials and birds). The male voids faeces and urine through its cloaca but has a penis for the transmission of sperm. In most other respects, monotremes are typical mammals. They have a covering of hair, are warm-blooded and suckle their young on milk which is secreted through numerous pores on the belly (not through teats, as in other mammals). The male Platypus is unique among mammals in being venomous; toxin is injected through a hollow spur on the ankle.

The order comprises two families (possibly better regarded as superfamilies or suborders), the Ornithorhynchidae and Tachyglossidae.

Family ORNITHORHYNCHIDAE

(or'-nith-oh-rink'-id-ee: "*Ornithorhynchus*-family")

This family includes one living species, the Platypus. A fossil species resembles it in general body shape but differs in having teeth.

Genus Ornithorhynchus

(or'-nith-oh-rink'-us: "bird-snout")

The characteristics of the genus are those of the species.

Platypus

Ornithorhynchus anatinus (an'-ah-tee'-nus: "duck-like bird-snout")

(RH Green)

The Platypus feeds on aquatic insect larvae, shrimps and worms by dabbling in mud or silt on the bottom of rivers and freshwater lakes with its sensitive, flexible, duck-like beak. In addition to the sense of sound, the beak is sensitive to slight electric currents generated by prey. Its eyes are shut while it is under water and, being buoyant, it must continuously swim downward with its webbed forefeet to remain submerged. The hind feet are also webbed, but are employed in steering or braking—not in propulsion. It usually feeds at night, with peaks of activity for several hours after dusk and before dawn. During the day, it rests in a burrow in the bank of the river or lake, but it may spend some hours near the entrance to the burrow, basking in the sun and grooming its dense fur.

The male has a sharp, hollow, horny spur on the inside of the ankle. This is connected to a venom gland which produces a very strong toxin. The spur can be used in defence against predators, but the fact that it is restricted to the male—and that the gland reaches its greatest development in the mating season—suggests that it is normally employed in aggressive encounters between males.

Males are larger than females. Mating occurs once a year, the season beginning in August in the warmer northern parts of the range and in October in the southern part. The female usually lays two eggs and incubates these against her abdomen for about two weeks in a blocked-off nest at the end of a long breeding burrow. One or two young are suckled for four to five months.

HABITAT: edges of subtropical to cool temperate rivers and freshwater lakes where burrows can be dug
HEAD AND BODY: 30–42 cm
TAIL: 10–13 cm
DISTRIBUTION: 300,000–1 million km²
ABUNDANCE: sparse
STATUS: probably secure

Family TACHYGLOSSIDAE

(tak'-ee-glos'-id-ee: "*Tachyglossus*-family")

Members of this family have spines of varying length interspersed with fur on the back, sides and tail. The snout is tubular with a very small mouth, through which the long tongue can be rapidly extruded and retracted. The Short-beaked Echidna, which occurs in Australia and the lowlands of New Guinea, feeds mainly upon ants, which adhere to its sticky tongue. The much larger Long-beaked Echidna, restricted to rainforest (or mist forest) in the highlands of New Guinea, feeds mainly on worms that it finds in the litter of the forest floor—apparently grasping these with the tip of its tongue.

Genus Tachyglossus
(tak'-ee-glos'-us: "fast-tongue")

The single species of this genus is smaller than the Long-beaked Echidna and has a tongue without grasping structures at its tip.

Short-beaked Echidna

Tachyglossus aculeatus (ak-yue'-lay-ah'-tus: "spiny fast-tongue")

The Short-beaked Echidna (usually referred to in Australia simply as the Echidna) spends most of the day sleeping in the shelter of fallen timber or rocks. It is able to sink itself below the surface of the soil without moving forwards or backwards and offers very great resistance to dislodgement once it has wedged itself under cover. Depending upon the temperature and the availability of food, it feeds for varying lengths of time between dusk and dawn on ants or termites, breaching their nests with its forefeet and licking up the insects with its sticky tongue. The male has a spur on its ankle, similar to that of a Platypus, but it is not hollow and is not connected to a venom gland.

The species occurs over most of Australia, including Tasmania, and in the lowlands of New Guinea. It varies considerably in appearance: animals from the cooler part of the range have hair almost as long as the spines; in those from the hotter, drier parts, the hair is barely visible between the spines.

Mating occurs in July and August. The single egg is laid into a temporary pouch which develops at the onset of breeding by the outgrowth of folds of skin on the abdomen of the female, and it hatches after about 10 days. The young is suckled in the pouch until the spines begin to develop. Still blind, it is then left in a short burrow while the mother is feeding. It is suckled for at least three months.

HABITAT: cool temperate to tropical wet sclerophyll forest to desert
HEAD AND BODY: 37–50 cm
TAIL: 8–9 cm
DISTRIBUTION: more than 1 million km^2
ABUNDANCE: common
STATUS: secure

(GB Baker)

Subclass MARSUPIALIA
(mar-sue'-pee-ah'-lee-ah: "pouched [animals]")

Various anatomical features can be used to define marsupials but the most obvious is that the young are born with their hind legs in an embryonic condition. Since newly-hatched monotremes have similarly undeveloped hind limbs, it is necessary to emphasise that marsupials are *born* in this condition.

A newborn marsupial uses its precociously developed forelimbs to clamber through its mother's fur to a teat. It clamps onto this firmly with its mouth and remains attached until its eyes open, its fur grows and it is capable of independent movement. In most marsupials the teats lie inside a pouch on the mother's belly, which provides warmth, humidity and physical protection for the developing young. The opening of the pouch may be directed backwards (probably the primitive arrangement) or downwards, but a forward-opening pouch is found only in possums and kangaroos. Since the first marsupials known to European scientists were all equipped with pouches, it was assumed that this "external womb" was an absolute characteristic of marsupials. However, some species lack this structure; in some it is rudimentary; in others it develops only during the breeding season. Thus, the name Marsupialia refers to a common but not universal condition.

The widespread idea that marsupials are "inferior" or "primitive" mammals is not well-founded. Fossil evidence indicates that marsupials have had a separate evolutionary history of about the same age as the more familiar eutherian mammals and they are certainly not ancestral to these. However, in general, marsupials have smaller brains than eutherian mammals of corresponding size and they also have a lower rate of metabolism. An interesting feature of marsupials is that the first toe of the hind foot lacks a claw and, typically, can be opposed to the other four toes to grip a branch when climbing. This digit is best developed in climbing marsupials but is reduced (or even absent) in many terrestrial species.

Although fossil marsupials up to 100 million years old are known from the Americas and Europe, living marsupials are restricted to the Australian region (including Melanesia) and the South American continent. The only exception is the very adaptable Virginian Opossum, which occurs over much of North America. There are many significant differences between American and Australian marsupials and their relationships are not clearly understood. For a long time it was thought that, because the South American marsupials, the *carnivorous* Australian marsupials and the bandicoots had a rather similar arrangement of teeth (three or four pairs of slender upper incisors and three pairs in the lower jaw), all three should be linked together in a major group, the Polyprotodonta ("many-front-teeth"). The possums and kangaroos, which usually have three pairs of broad upper incisors and never more than one pair of *functional* incisors in the lower jaw, were placed in an alternative group, the Diprotodonta.

Over the past decade or so it has become increasingly difficult to defend this concept, for it does not correspond to any reasonable reconstruction of the evolutionary history of the marsupials. On the present evidence from fossils, comparative anatomy and biochemistry, the best hypothesis is that all the Australian marsupials have a common ancestry which is separate from that (or those) of the South American marsupials. The two groups are therefore placed respectively in the cohorts Australidelphia and Ameridelphia—but with a slight complication. A single, small species, *Dromiciops australis*, from Chile and western Argentina, is placed in the Australidelphia by some experts.

Cohort AUSTRALIDELPHIA

(os-trah'-lee-del'-fee-ah: "Australian marsupials")

This group, which includes all of the non-American marsupials, cannot be rigidly defined on significant anatomical characters but is strongly supported by immunological evidence. One significant difference between the Australidelphia and the Ameridelphia appears to be that the sperms of the former are single (as in most mammals) whereas those of the latter are paired (head joined and tails separate).

The cohort is divided into four readily recognisable orders. The Dasyuromorphia includes the primarily carnivorous and insectivorous species; the Peramelomorphia contains the omnivorous bandicoots and bilbies; the Diprotodontia includes the basically herbivorous possums and kangaroos, and the Notoryctemorphia is restricted to the very peculiar Marsupial Mole.

Opinions on the relationships of these orders to each other have varied considerably. Many zoologists have linked the Peramelomorphia with the Dasyuromorphia because of similarities in their dentition; others regard the Peramelomorphia as allied to the Diprotodontia because of a similarity of the hind feet. The Notoryctemorphia has been regarded by some scholars as most clearly related to the Dasyuromorphia, by others to the Diprotodontia.

Regarding these questions as insoluble on present evidence, we shall avoid any expression of opinion and treat each order quite separately.

Order DASYUROMORPHIA

(daz'-ee-yue'-ree-mor'-fee-ah: "*Dasyurus*-form-order")

This group includes predatory marsupials ranging from the size of a dog to some which could fit comfortably into a matchbox. The members of the order have four pairs of needle-like incisors in the upper jaw and three similar pairs in the lower jaw; the canine teeth are well developed and the molars and premolars are sharply serrated. (An exception to this generalisation is the Numbat, in which all the teeth are reduced to small, similar pegs.) The snout is usually elongate and the legs of similar length. The tail is never prehensile.

Family DASYURIDAE

(daz'-ee-yue'-rid-ee: "*Dasyurus*-family")

The characteristics of this group are essentially those of the superfamily. With the exception of the Tasmanian Devil, which is a terrier-sized scavenger, dasyurids are medium-sized to small predators, most of which are terrestrial although a few also hunt in trees. There are four subfamilies: Dasyurinae, Phascogalinae, Planigalinae and Sminthopsinae.

Subfamily DASYURINAE
(daz'-ee-yue'-reen-ee: "*Dasyurus*-subfamily")

This group, defined largely on biochemical criteria, includes the quolls (which are here regarded as belonging to the single genus *Dasyurus*); the dibblers, pseudantechinuses and kalutas (all of which were once regarded as antechinuses); and the Tasmanian Devil.

Genus *Dasycercus*
(daz'-ee-ser'-kus: "hairy-tail")

The single species of this genus is a terrestrial dasyurid with a black crest along the posterior part of the tail and five toes on the hind foot. The third premolars are small, if present, in the upper jaw and absent from the lower jaw. The female does not have a pouch, merely a ridge of skin surrounding the mammary area.

Mulgara
Dasycercus cristicauda (kris'-tee-kaw'-dah: "crest-tailed hairy-tail")

(*P Woolley, D Walsh*)

In his monograph on the Mulgara, Frederick Wood Jones referred to it as a "generalised marsupial". With increased knowledge of marsupials we now see that, while it is anatomically unspecialised, the Mulgara is, physiologically, one of the most perfectly adapted of the desert marsupials. It burrows by day and, at night, is an efficient predator on large arthropods and small rodents.

Sexual maturity is reached in the first year. Most mating takes place in March and April but is dependent upon rainfall. The female has up to eight teats on which the young are carried for about seven weeks. After weaning, young are suckled for a further seven weeks.

HABITAT: sandy inland deserts
HEAD AND BODY: 12–22 cm
TAIL: 7–13 cm
DISTRIBUTION: more than 1 million km^2
ABUNDANCE: sparse
STATUS: secure

Genus *Dasykaluta*
(daz'-ee-kah-lue'-tah: "kaluta–[in family]–Dasyuridae"; kaluta is an Aboriginal [Nyamal] name for the only known species in this genus)

The single species is very similar externally to *Antechinus* but its tail is thicker at base and the upper jaw lacks a third premolar. The female does not have a pouch, merely a ridge of skin surrounding the mammary area.

Little Red Kaluta

Dasykaluta rosamondae (roz'-ah-mon'-dee: "Rosamond-related kaluta-dasyurid", referring to red-haired Rosamond, mistress of Henry II)

Until recently, this species was classified in the genus *Antechinus*, which it resembles in appearance. Even after its transfer to a genus of its own (*Dasykaluta*), it has been referred to as the Little Red Antechinus—which is confusing. It is here called a kaluta, this being an Aboriginal name for the species. For the sake of continuity, the name is expanded to Little Red Kaluta. Its tail, which is thickened at the base, is flicked in the air in an unusual manner when the animal is feeding. It feeds on insects and small lizards, mainly at night.

Males are much larger than females. Sexual maturity is reached at about 11 months. Mating takes

(P Woolley, D Walsh)

place in September or October. The female has eight teats and may carry up to eight young. These attain independence at about 16 weeks. Males die shortly after mating.

HABITAT: subtropical arid spinifex tussock grassland
HEAD AND BODY: 9–11 cm
TAIL: 6–7 cm
DISTRIBUTION: 100,000–300,000 km²
ABUNDANCE: sparse
STATUS: probably secure

Genus *Dasyuroides*

(daz'-ee-yue-roy'-daze: "*Dasyurus*-like [animal]")

The single species of *Dasyuroides* is a terrestrial dasyurid with a black brush on the posterior part of the tail and four toes on the hind feet. The third premolars are very small and may be absent from the lower jaw. The female does not have a pouch, only a ridge of skin on either side of the mammary area.

Kowari

Dasyuroides byrnei (ber'-nee: "Byrne's *Dasyurus*-like animal", after P. M. Byrne, who collected the first specimen)

Seldom seen in the wild, this attractive dasyurid is now reasonably familiar to people who visit the nocturnal houses of the major Australian zoos. It is well adapted to life in the central desert and does not need to drink; when cold, it may become torpid. By day, it sleeps in a burrow (sometimes emerging to bask in the sun), and at night it is

a fierce predator on small vertebrates and the larger arthropods. Its range seems to have contracted considerably in recent decades but it is not clear whether this is an indication of its impending endangerment or of cyclical changes in the density of an opportunistic species.

Sexual maturity is reached in the first year of life but breeding seldom

takes place until the second year. The female carries up to six young on her teats for about eight weeks and suckles them in a nest for a further eight weeks.

HABITAT: subtropical to tropical stony desert
HEAD AND BODY: 13–18 cm
TAIL: 11–14 cm
DISTRIBUTION: 100,000–300,000 km²
ABUNDANCE: sparse
STATUS: vulnerable

(K Johnson)

Genus *Dasyurus*

(daz'-ee-yue'-rus: "hairy-tail")

The body in this genus is long and the snout rather pointed. The fur of the body is white-spotted, but only in one species (*D. maculatus*) do the spots continue into the long furry tail. There are two pairs of premolar teeth in the upper and lower jaws. The canine teeth are well developed. Females have shallow pouches, surrounding six teats.

Western Quoll

Dasyurus geoffroii (zhe-froy'-ee: "Geoffroy's hairy-tail", after French zoologist E. Geoffroy Saint-Hilaire)

Similar in size to the Eastern Quoll, this species is well adapted to high temperatures and aridity. These features once permitted it to occupy the greater part of inland Australia but, since European settlement, its range has diminished to the relatively well watered south-western corner of the mainland. Its decline may have been due to an alteration of the way of life of Aborigines, who were prevented from continuing their regular burning of the understorey of open forests and savanna—a practice which led to a flush of re-generating vegetation (and insects, etc.) and provided an optimal habitat. It is an opportunistic predator on small vertebrates and large insects, taken on the ground or in trees. The Western Quoll differs from the Eastern Quoll in possessing a (small) first toe on the hind foot.

(J Lochman)

Sexual maturity is reached at about one year. Up to six young are suckled on six teats for about 15 weeks.

HABITAT: desert to wet sclerophyll forest
HEAD AND BODY: 27–36 cm
TAIL: 20–28 cm
DISTRIBUTION: 30,000–100,000 km²
ABUNDANCE: sparse
STATUS: vulnerable

Northern Quoll

Dasyurus hallucatus (hal'-uke-ah'-tus: "first-toed hairy-tail")

This is the smallest and, at present, the most successful of the quolls. Over its extensive northern range it is common in the wild and a minor pest around settlements, entering houses in search of food scraps. Its natural diet comprises small vertebrates, large insects and fruits.

Sexual maturity is reached at about one year and mating takes place in June and July. The female has six or eight teats but lacks a pouch. About six young are suckled on the teats for about 10 weeks and subsequently in a nest for another 10 weeks.

Like the Western Quoll, the Northern Quoll has a first digit on the hind foot but it is not notably arboreal.

HABITAT: tropical woodland, particularly in rocky country
HEAD AND BODY: 12–31 cm
TAIL: 20–30 cm
DISTRIBUTION: more than 1 million km²
ABUNDANCE: common
STATUS: secure

(J Lochman)

Spotted-tailed Quoll

Dasyurus maculatus (mak-yue-lah'-tus: "spotted hairy-tail")

Largest of the quolls, this species is the only one in which the pattern of spots on the body is continued into the tail. The background colour of the fur is usually a rich russet brown, but some individuals are chocolate or even black. The long tail is not prehensile. Primarily a predator on a wide range of terrestrial and arboreal vertebrates, it also eats carrion. It is nocturnal, sheltering by day in a tree-hole, hollow log or rock crevice, but it may bask or forage in daylight when the weather is cold.

Sexual maturity is reached at about one year. Mating extends from April to July and up to six young are suckled on six teats in a shallow pouch. Young leave the pouch when about seven weeks old and are suckled in a nest for a further six weeks.

HABITAT: wet and dry sclerophyll forest, rainforest
HEAD AND BODY: 35–76 cm
TAIL: 34–55 cm
DISTRIBUTION: 300,000–1 million km²
ABUNDANCE: sparse
STATUS: probably secure

(JE Wapstra)

Eastern Quoll

Dasyurus viverrinus (viv'-er-een'-us: "ferret-like hairy-tail")

(GB Baker)

Sexual maturity is reached at about one year of age and mating occurs from May to June. Up to six young are suckled on six teats for about six weeks and thereafter for about the same time in a nest.

HABITAT: wet and dry sclerophyll forest, particularly where abutting on grassland or pasture; heathland
HEAD AND BODY: 28–45 cm
TAIL: 17–28 cm
DISTRIBUTION: 30,000–100,000 km²
ABUNDANCE: sparse
STATUS: probably secure

Once occupying most of the coastal forests of south-eastern Australia, the Eastern Quoll is now restricted to Tasmania, where its range coincides with that of the much larger Spotted-tailed Quoll. It is an opportunistic predator on small vertebrates and large insects and their larvae; it also eats carrion, seeds and fruits. It is nocturnal, sleeping by day in or under logs or in similar ground shelter: it may become torpid for short periods in cold weather. Black-furred individuals are not uncommon. The hind foot has only four toes.

Genus *Parantechinus*

(pa'-ran-te-kie'-nus: "alongside-*antechinus*")

Originally, Dibbler was a name restricted to a species in the genus *Antechinus*. With the removal of this species to a newer genus *Parantechinus* and the addition of another erstwhile *Antechinus* species to this genus, it becomes convenient and meaningful to refer to these as Northern and Southern Dibblers.

Dibblers resemble antechinuses in appearance, but the second upper incisor is about equal to, or shorter than, the fourth, and the upper and lower third premolars are shorter than the second. Females either lack a pouch or have a very poorly developed one.

Southern Dibbler

Parantechinus apicalis (ah'-pik-ah'-lis: "pointed alongside-*antechinus*", referring to tapering tail)

This very rare species sleeps by day in a nest of twigs and grass. At night it burrows through leaf litter in search of insects or climbs into low trees, particularly banksias, to sip nectar from flowers and possibly to eat insects that gather at the blossoms.

Males are noticeably larger than females. Sexual maturity is attained at about 11 months. Mating takes place around March. The female has eight teats in a very shallow pouch and usually carries eight young which become independent at about 16 weeks. Males do *not* die shortly after the first mating.

HABITAT: cool-temperate banksia heathland
HEAD AND BODY: 12–14 cm
TAIL: 9–11 cm
DISTRIBUTION: less than 10,000 km²
ABUNDANCE: rare
STATUS: endangered

(R Whitford)

Northern Dibbler

Parantechinus bilarni (bil-ar'-nee: "Bill Harney's alongside-*antechinus*", after William (Bill) Harney, guide to expedition which collected first specimen)

(P Woolley, D Walsh)

The Norther Dibbler, until recently known as the Sandstone Antechinus, inhabits rocky screes. It sleeps during the day in a crevice among the rocks and forages at night for insects, on the ground and in trees. In the dry season some animals move from the screes into nearby vine-thickets which provide less arid surroundings.

Males are only slightly larger than females. Sexual maturity is reached at about 11 months and mating occurs from June to August. The female has six teats but no pouch. The usual litter is four or five young which become independent at 16 to 20 weeks. There is no mass death of males after breeding: some breed again in their second year.

HABITAT: rock screes in tropical sandstone country with abundant vegetation
HEAD AND BODY: 8–10 cm
TAIL: 10–12 cm
DISTRIBUTION: 30,000–100,000 km²
ABUNDANCE: common
STATUS: probably secure

Genus *Pseudantechinus*

(sude'-an-te-kie'-nus: "false-*antechinus*")

Pseudantechinus is similar to antechinuses in appearance but has a much shorter tail, swollen at the base. As in *Parantechinus*, the upper third premolar is very small: it is absent from the lower jaw.

Fat-tailed Pseudantechinus

Pseudantechinus macdonnellensis (mak-don'-el-en'-sis: "Macdonnell [Range] false-*antechinus*")

Of all the antechinus-like marsupials, this species occupies the most arid habitat, usually in rocky country. As the common name indicates, the tail is thick at its base and is markedly tapered: fat is deposited in the tail when food—mostly insects—is abundant.

Sexual maturity is reached at about 11 months. Mating may occur around June or August, but females breed only once a year. There are six teats in a moderately developed pouch, and a litter may comprise up to six young. Males do not die after mating and usually survive to breed in their second year.

HABITAT: tropical to subtropical rocky, arid woodland or grassland
HEAD AND BODY: 9–11 cm
TAIL: 7–9 cm
DISTRIBUTION: more than 1 million km²
ABUNDANCE: common
STATUS: secure

(*P Woolley, D Walsh*)

Ningbing Pseudantechinus

Pseudantechinus ningbing (ning'-bing: "Ningbing false-*antechinus*", after Ningbing Station, Kimberley, where first specimen was taken)

HABITAT: tropical woodland and shrubland on deeply dissected sandstone
HEAD AND BODY: 7–9 cm
TAIL: 7–9 cm
DISTRIBUTION: 30,000–100,000 km²
ABUNDANCE: sparse
STATUS: probably secure

(*P Woolley, D Walsh*)

Like the Fat-tailed Antechinus, this species has a tail which is markedly thicker at the base and, also like that species, it inhabits rocky country. Little is known of its biology.

Males are larger than females.

Sexual maturity is reached at about 11 months. The female lacks a pouch and has four teats. It seems that males do not die after mating and that some survive to breed in the second year of life.

Western Pseudantechinus

Pseudantechinus woolleyae (wool'-ee-ee: "Woolley's false-*antechinus*", after P. A. Woolley, Australian zoologist)

This species resembles the other pseudantechinuses in general appearance but is the largest of these. It differs from the Ningbing Pseudantechinus in having six (rather than four) teats, and a tail which is shorter than the head and body (about the same length in the Ningbing Pseudantechinus).

In well-nourished individuals, the tail becomes carrot-shaped due to storage of fat.

HABITAT: warm temperate arid to semi-arid rocky country, often with mulga and hummock grasses
HEAD AND BODY: 8–10 cm
TAIL LENGTH: 7–9 cm
DISTRIBUTION: 300,000–1 million km²
ABUNDANCE: sparse
STATUS: probably secure

Genus *Sarcophilus*

(sar-kof'-il-us: "flesh-lover")

The single species in this genus is unmistakable. The body is short and the head is large, with rather short, powerful jaws. The black fur usually has some white patches. Individuals may weigh up to 12 kilograms.

Tasmanian Devil

Sarcophilus harrisii (ha'-ris-ee-ee: "Harris's flesh-lover", after G. P. R. Harris, who described the species)

Largest of the surviving dasyurids, the Tasmanian Devil is a scavenger which also forages upon insect larvae and will attack penned poultry and weak or juvenile wallabies and sheep. It is normally active at night, sleeping by day under ground cover.

Sexual maturity is reached at the age of about two years and mating takes place in March and April. The female has four teats in a well-developed backward-directed pouch and normally suckles two or three young in the pouch for about 15 weeks and in a den for a further 15 weeks.

Numbers have varied widely. Since the 1960s, the species has been very common.

HABITAT: wet sclerophyll forest to woodland
HEAD AND BODY: 50–65 cm
TAIL: 23–26 cm
DISTRIBUTION: 30,000–100,000 km²
ABUNDANCE: common
STATUS: secure

(GB Baker)

Subfamily PHASCOGALINAE

(fas'-koh-gah'-ee'-nee: "*Phascogale*-subfamily")

This group, defined largely on biochemical criteria, includes the phascogales and antechinuses.

Genus *Antechinus*

(an'-te-kie'-nus: "hedgehog-equivalent")

The generic name refers to the somewhat spiky fur of the Brown Antechinus and to the initial belief of the namer of the genus that it was, like hedgehogs, a member of the order Insectivora. The genus extends into New Guinea, where there are at least four species.

Antechinuses are somewhat similar in appearance to rats or mice but have a more acutely tapering muzzle, shorter ears and a fully haired tail. The hind feet are short and broad, retain a small first digit and have transversely striated pads. All of these features appear to be arboreal adaptations, but most antechinuses are mainly terrestrial. Males are larger than females and usually die within a few weeks of mating. Females have six to 12 teats in a very rudimentary pouch.

Fawn Antechinus

Antechinus bellus (bel'-us: "beautiful hedgehog-equivalent")

Unlike the other two tropical antechinuses (*A. godmani* and *A. leo*), which inhabit rainforest, this species lives in open forest and woodland where leaf litter is sparse and dry. Not surprisingly, therefore, it seeks insect food in trees as well as on the ground. It nests during the day in a tree-hole or a fallen hollow log.

Males are markedly larger than females. Mating occurs in August or September. The female has 10 teats and carries up to 10 young. Males die shortly after mating.

HABITAT: tropical dry eucalypt forest to woodland with abundant tree-holes or hollow logs
HEAD AND BODY: 11–15 cm
TAIL: 9–13 cm
DISTRIBUTION: 100,000–300,000 km²
ABUNDANCE: abundant
STATUS: probably secure

(*P Woolley, D Walsh*)

Yellow-footed Antechinus

Antechinus flavipes (flah'-vee-pez: "yellow-footed hedgehog-equivalent")

One of the most successful of the middle-sized dasyurids, the Yellow-footed Antechinus is a common species over most of non-arid eastern Australia and is represented by a sub-

species in south-western Australia. It is an adaptable animal which, unlike most dasyurids, invades households in search of mice. It sleeps by day in a nest under ground cover and hunts at night for large arthropods and very small terrestrial vertebrates: it also feeds on flowers and nectar.

Males are notably larger than females. Sexual maturity is attained at about 12 months and mating takes place in August and September. Males die shortly after copulation. The female lacks a pouch but develops ridges around the mammary area before giving birth. There are usually 10 teats (varying from eight to 12) and it is common for one young to be carried on each teat for about five weeks, after which they are suckled in a nest for up to 15 weeks.

(R Whitford)

HABITAT: tropical rainforest to cool-temperate sclerophyll forest and woodland
HEAD AND BODY: 9–17 cm
TAIL: 6–15 cm
DISTRIBUTION: more than 1 million km²
ABUNDANCE: abundant
STATUS: secure

Atherton Antechinus

Antechinus godmani (god'-mun-ee: "Godman's hedgehog-equivalent", after F. D. Godman, husband of sponsor of expedition which discovered this species)

(R Whitford)

has six teats in a rudimentary pouch. Litters comprise up to six young which are carried for about five weeks and thereafter suckled in a nest for an unknown period. Males die soon after mating.

HABITAT: very wet montane tropical rainforest (mist forest)
HEAD AND BODY: 9–16 cm
TAIL: 9–14 cm
DISTRIBUTION: less than 10,000 km²
ABUNDANCE: rare
STATUS: vulnerable

Largest of the antechinuses, this species was not described until 1982. Like the Cinnamon Antechinus, it has an extremely restricted range in mist forest in the vicinity of Ravenshoe, Queensland. It feeds at night on insects in the leaf litter and

decomposing timber of the forest floor, locating its prey by scent and digging it out with its forefeet.

Males are markedly larger than females. Sexual maturity is reached at about 11 months. Mating takes place in July or August. The female

Cinnamon Antechinus

Antechinus leo (lay'-oh: "lion hedgehog-equivalent", from Leo Creek, McIlwraith Range, Queensland; also refers to "lion-like" colour)

Northernmost of the east coast antechinuses, this species occupies a very restricted area of vine-forest in the vicinity of Iron Range on Cape York Peninsula. It is an arboreal insectivore which probably also forages on the ground. Little is known of its biology.

Males are notably larger than females. Mating probably takes place in October. The female has 10 teats.

HABITAT: tropical vine-forest
HEAD AND BODY: 9–16 cm
TAIL: 8–14 cm
DISTRIBUTION: less than 10,000 km²
ABUNDANCE: common
STATUS: vulnerable

(P Woolley, D Walsh)

Swamp Antechinus

Antechinus minimus (min'-im-us: "smallest hedgehog-equivalent")

Despite its scientific name, this is one of the larger antechinuses. The anomaly is due to its having been described originally as a species of the genus *Dasyurus* (of which it was the smallest member). It is the southernmost of the antechinuses, being restricted to Tasmania, Bass Strait islands and southern Victoria and South Australia, where it inhabits dense heathland or tussock grassland. It feeds, like the Dusky Antechinus, by digging in the soil for shallow-burrowing insects and other arthropods.

Males are larger than females.

Sexual maturity is reached at about 11 months and mating occurs from June to August. All males die after mating. Females have six teats (Tasmania) or eight (mainland) and usually carry a young on each teat. Subsequently they are suckled in a nest.

HABITAT: cool-temperate to cold wet heathland, tussock grassland or sedgeland
HEAD AND BODY: 9–14 cm
TAIL: 6–10 cm
DISTRIBUTION: 100,000–300,000 km²
ABUNDANCE: very sparse
STATUS: probably secure

(D Milledge)

Brown Antechinus

Antechinus stuartii (styue'-ar-tee-ee: "Stuart's hedgehog-equivalent", after J. Stuart, who made a preliminary description of the species)

The distribution of this species lies within that of the Yellow-footed Antechinus. It is absent from Western Australia and, in eastern Australia, is largely restricted to wet sclerophyll forests on the coastal side of the Dividing Range. It feeds on large insects and other arthropods, mostly in forest litter; in the drier parts of its range it climbs trees in search of food. It also tends to be more arboreal in areas where the Yellow-footed Antechinus is numerous. During the day it sleeps in a nest under shelter.

Males are notably larger than females. Sexual maturity is attained at about 12 months and mating occurs in August and September. Males die shortly after mating. Females have a rudimentary pouch surrounding six to 10 teats. Young are carried on the teats for about five weeks, after which they are suckled in a nest for up to 15 weeks.

(CA Henley)

HABITAT: tropical to cool-temperate wet and dry sclerophyll forest
HEAD AND BODY: 7–14 cm
TAIL: 6–11 cm
DISTRIBUTION: 300,000–1 million km^2
ABUNDANCE: abundant
STATUS: secure

Dusky Antechinus

Antechinus swainsonii (swane'-sun-ee-ee: "Swainson's hedgehog-equivalent", after W. Swainson, British naturalist)

The Dusky Antechinus has a distribution similar to that of the Brown Antechinus but favours higher altitudes. It extends into Tasmania but does not occur in the tropics. Although mainly nocturnal, it often forages during the day for insects and other invertebrates that live just below the surface of the soil: these are located by scent and dug out with the forefeet. The diet may be supplemented by fruits and berries.

Males are noticeably larger than females. Sexual maturity is reached at about 11 months. Mating takes place in June or July, after which all males die. The female has a rudimentary pouch surrounding six to 10 teats and usually carries young on each teat for seven to eight weeks. These are subsequently suckled in a subterranean nest for about four weeks.

HABITAT: wet eucalypt forest with a dense understorey; alpine heathland
HEAD AND BODY: 9–19 cm
TAIL: 7–12 cm
DISTRIBUTION: 300,000– 1 million km^2
ABUNDANCE: abundant
STATUS: secure

(R Whitford)

Genus *Phascogale*

(fas'-koh-gah'-lay: "pouched-weasel")

The two species of this genus are arboreal dasyurids with a conspicuous brush on the posterior part of the tail. The lower third premolar is notably smaller than the second. The pouch is either absent or poorly developed.

Red-tailed Phascogale

Phascogale calura (kal-ue'-rah: "beautiful-tailed pouched-weasel")

The Red-tailed Phascogale, which is an arid-adapted sister-species of the Brush-tailed Phascogale, was widely distributed throughout inland Australia at the time of European settlement. For unknown reasons, but perhaps in response to competition from the feral Cat, it is now restricted to small populations in south-western Australia, where its survival may be related to the presence of "poison weeds" (the peas *Gastrolobium* and *Oxylobium*). Like the Brush-tailed Phascogale, it is an agile climber, but it finds most of its food—small vertebrates and larger arthropods—on the ground.

Sexual maturity is reached at about 12 months and mating takes place in May and June. Up to eight young are reared. It is probable that males die shortly after mating.

HABITAT: semi-arid eucalypt forest with continuous canopy
HEAD AND BODY: 9–13 cm
TAIL: 12–15 cm
DISTRIBUTION: 10,000–30,000 km²
ABUNDANCE: sparse
STATUS: possibly endangered

(AG Wells)

18

Brush-tailed Phascogale

Phascogale tapoatafa (tap'-oh-ah-tah'-fah: "pouched-weasel [known to Aboriginals as] tapoa-tafa")

(H & J Beste)

In trees, the Brush-tailed Phascogale is by far the most agile of the semi-arboreal dasyurids, but its brush-tipped tail is not prehensile. By day it rests in a leaf-lined nest in a tree-hole and at night it preys upon small arboreal mammals, nestling birds and larger arthropods. It also feeds on the ground and occasionally attacks penned poultry.

Sexual maturity is reached at the age of about one year and mating occurs around June; males usually die within a week or so after copulating. The female, which lacks a pouch, usually carries three to six young on her eight teats for about six weeks. They are suckled in a nest for a further 14 weeks. Juveniles may remain in the maternal nest until just before reaching sexual maturity.

HABITAT: open dry sclerophyll forest on ridges and rocky slopes with little ground cover
HEAD AND BODY: 16–23 cm
TAIL: 17–22 cm
DISTRIBUTION: 300,000–1 million km²
ABUNDANCE: sparse
STATUS: probably secure

Subfamily PLANIGALINAE

(plan'-i-gah-leen'-ee: "*Planigale*-subfamily")

This group, defined largely on biochemical criteria, includes the planigales and ningauis.

Genus Ningaui

(nin-gow'-ee: "ningaui", the name of an Aboriginal mythical being which is said to be tiny, nocturnal, hairy and short-footed and to eat raw flesh)

The Aboriginal name given to this genus reflects the small size of these marsupial predators and the fact that their hind feet are relatively shorter than those of dunnarts. Ningauis look rather like planigales but the head is less flattened and the fur rather more bristly. The pouch is either absent or poorly developed.

Wongai Ningaui

Ningaui ridei (rie'-dee: "Ride's *ningaui*", after W. D. L. Ride, Australian zoologist)

(HJ Aslin)

This species inhabits sandy deserts, usually dunes with spinifex hummocks. It preys on a wide variety of insects found on the ground and among the leaves of vegetation, climbing with the help of its weakly prehensile tail. During the day it sleeps in the base of a hummock, under fallen timber or in a short burrow.

Mating occurs from September to January. The female has six or seven teats in a rudimentary pouch and may carry up to seven young, which quit the pouch when about six weeks old and become independent at about 13 weeks. Two litters may be reared in a year.

HABITAT: subtropical to warm-temperate arid to semi-arid dunes with spinifex hummocks or shrub cover
HEAD AND BODY: 6–7 cm
TAIL: 6–7 cm
DISTRIBUTION: 300,000–1 million km²
ABUNDANCE: common
STATUS: secure

Pilbara Ningaui

Ningaui timealeyi (tim-ee'-lee-ee: "Tim Ealey's *ningaui*", after E. H. M. (Tim) Ealey, Australian zoologist)

Although the average weight of adult Pilbara Ningauis is greater than that of Long-tailed Planigales, the former could perhaps lay claim to being the smallest marsupial, inasmuch as independently feeding juveniles can weigh as little as two grams. The Pilbara Ningaui feeds at night on grasshoppers, cockroaches and centipedes. During the day, it sleeps in a nest in the base of a spinifex hummock or similar shelter.

Breeding takes place from September to March in good seasons, but from November to January when rain and food are restricted. The female has four to six teats but lacks a definite pouch. Up to six young may be reared, being carried on the teats for about six weeks and becoming independent at about 13 weeks. Few adults survive to breed in a second year.

HABITAT: subtropical semiarid shrubland and hummock grassland
HEAD AND BODY: 4–6 cm
TAIL: 5–8 cm
DISTRIBUTION: 100,000–300,000 km²
ABUNDANCE: common
STATUS: probably secure

(R Whitford)

Southern Ningaui

Ningaui yvonneae (ee-von'-ee: "Yvonne's *ningaui*", after Yvonne C. Kitchener, wife of D. J. Kitchener, Australian zoologist)

Little is known of the biology of this ningaui. It is active at night, probably as a predator upon insects and other invertebrates.

The female has seven teats and probably breeds in spring.

HABITAT: arid temperate sand plains and dunes with spinifix and mallee
HEAD AND BODY: 5–7 cm
TAIL: 6–7 cm
DISTRIBUTION: 300,000–1 million km²
ABUNDANCE: sparse
STATUS: secure

(CA Henley)

Genus *Planigale*
(plan'-ee-gah'-lay: "flat-weasel")

These very small carnivorous marsupials, related to the ningauis and dunnarts, are characterised by a flattened head which appears to be an adaptation for sheltering in narrow crevices, such as the cracks in sun-dried mud. The ears of planigales are shorter and the hind feet broader than in ningauis. The hind feet are turned outwards when the animal is running. During the breeding season, females have a well-developed pouch which opens to the rear and encloses eight to 12 teats: in some species the pouch regresses when females are not in a breeding condition.

Paucident Planigale

Planigale gilesi (jile'-zee: "Giles's flat-weasel", after Ernest Giles, explorer of Australian deserts)

(R Whitford)

The common name of this species refers to its having only two premolar teeth (other planigales have three). It hunts at night in leaf litter and on the stems of grasses for large and small insects; in cold weather it may forage for several hours during the day or, alternatively, it may conserve energy by becoming torpid. It shelters in cracks in the soil.

Breeding occurs from July to January, with a peak around September. The female has a well-developed pouch with 12 teats, but the average litter is six. Young quit the pouch at the age of five to six weeks and are then suckled in a nest to the age of about 11 weeks. Several litters may be reared in a year.

HABITAT: arid to semiarid river flats, channels and flood plains with grass, sedges or shrubs on clay soils that crack when dry
HEAD AND BODY: 6–8 cm
TAIL: 5–7 cm
DISTRIBUTION: 300,000–1 million km²
ABUNDANCE: sparse
STATUS: probably secure

Long-tailed Planigale

Planigale ingrami (in'-gram-ee: "Ingram's flat-weasel", after Sir William Ingram, sponsor of expedition which collected first specimens)

(P Woolley, D Walsh)

HABITAT: seasonally flooded tropical grassland
HEAD AND BODY: 5–6 cm
TAIL: 5–6 cm
DISTRIBUTION: 300,000–1 million km²
ABUNDANCE: very sparse
STATUS: probably secure

With an average weight of about 4 grams, this is the smallest marsupial and one of the smallest mammals in the world. Despite the common name, its tail is not remarkably elongate: it is usually slightly longer than the head and body, whereas in other planigales it is usually slightly less. It is a nocturnal predator on large insects and, during the day, sleeps in a nest in a natural crevice: in the dry season it is commonly found in the cracks of muddy soil.

Breeding occurs throughout the year but mainly in the wet season, from November to April. The female has eight to 12 teats in a well-developed pouch and rears four to 12 young in a litter. They quit the pouch when about six weeks old and are suckled to the age of about 12 weeks.

Common Planigale

Planigale maculata (mak'-yue-lah'-tah: "spotted flat-weasel")

Despite its scientific name, this species is very rarely spotted: its head, back and sides are usually a uniform grey-brown. With a weight of up to 22 grams, this is the largest of the planigales. It is a nocturnal predator on insects and small vertebrates that can be larger than itself. During the day it sleeps in a nest under a rock or fallen timber.

Males are larger than females. Females appear to be sexually mature at about eight months. In the northern parts of the range, breeding takes place throughout the year, with peaks in autumn and spring. In the southern part, mating is from October to January. The female has eight to 12 teats in a well-developed pouch: the average size of a litter is eight.

(R Whitford)

HABITAT: tropical to warm-temperate rainforest, sclerophyll forest, wooded grassland and marsh, usually with moist soil.
HEAD AND BODY: 7–10 cm
TAIL: 6–10 cm
DISTRIBUTION: 300,000–1 million km²
ABUNDANCE: common
STATUS: secure

Narrow-nosed Planigale

Planigale tenuirostris (ten'-ue-ee-ros'-tris: "slender-snouted flat-weasel")

Although its range extends into semiarid and desert country, the distribution of the Narrow-nosed Planigale is restricted to the vicinity of watercourses and lakes that may diminish or disappear during dry conditions but periodically overflow onto flood plains. It usually hunts by night, feeding on large and small insects, but in cold weather it may hunt during the day or reduce its energy consumption by becoming torpid.

Males are slightly larger than females. Breeding is known to occur from August to February. The female has a temporary pouch surrounding 10 or 12 teats, but usually rears about six young. Females may breed more than once a year.

HABITAT: warm-temperate dense grassland on flood plains of rivers and lakes in semiarid to arid country
HEAD AND BODY: 5–8 cm
TAIL: 5–7 cm
DISTRIBUTION: more than 1 million km²
ABUNDANCE: sparse
STATUS: probably secure

(R Whitford)

Subfamily SMINTHOPSINAE

(smin'-thop-seen'-ee: "*Sminthopsis*-subfamily")

This group, defined largely on biochemical criteria, includes the dunnarts and the Kultarr.

Genus Antechinomys

(an'-te-kie'-noh-mis: "*antechinus*-mouse")

On the basis of skull anatomy, the single species in this genus has recently been included in the genus *Sminthopsis*. It is retained here in its original genus because it differs very significantly from *Sminthopsis* in the great length of its hind legs and tail. The hind foot lacks the first digit.

Kultarr

Antechinomys laniger (lan'-i-jer: "wool-bearing *antechinus*-mouse")

In its head and body, the Kultarr is almost indistinguishable from a dunnart. It differs in having very long, slender hind legs and a very long, tufted tail. Such structures are usually associated with hopping, but the Kultarr bounds, somewhat like a rabbit. It is a nocturnal predator on insects and other arthropods. During the day it sleeps in a nest under a stone or log, in a crack in the soil or in a short burrow.

Males are slightly larger than females. Females become sexually mature at eight months. Mating takes place from June to October (perhaps until February), and two litters may be reared in a season. The female usually has six to eight teats (sometimes four or 10) in a rudimentary pouch and usually rears about four young in a litter. These detach from the teats at about four weeks and are suckled in a nest until about 13 weeks old. Some young may cling to the mother's fur while she is foraging.

HABITAT: arid to semiarid grassland, scrubland and stony desert
HEAD AND BODY: 7–11 cm
TAIL: 10–15 cm
DISTRIBUTION: more than 1 million km²
ABUNDANCE: sparse
STATUS: probably secure

(K Johnson)

Genus Sminthopsis

(smin-thop'-sis: "mouse-appearance")

Dunnarts are mouse-sized dasyurids with very pointed muzzles, large eyes and ears. The hind feet are proportionately much longer and narrower than in antechinuses. The third premolar teeth are almost as large as the second. It seems that in most (possibly all) species, the female has a well-developed, downward opening pouch, enclosing from six to ten teats.

Kangaroo Island Dunnart

Sminthopsis aitkeni (ayt'-ken-ee: "Aitken's mouse-appearance", after P. F. Aitken, Australian zoologist)

Nothing is known of the biology of this dunnart, which is restricted to Kangaroo Island.

HABITAT: cool-temperate mallee
HEAD AND BODY: 8–9 cm
TAIL: 9–10 cm
DISTRIBUTION: less than 10,000 km²
ABUNDANCE: very sparse
STATUS: vulnerable

Chestnut Dunnart

Sminthopsis archeri (ar'-cher-ee: "Archer's mouse-appearance", after M. Archer, Australian zoologist)

First described in 1986, the Chestnut Dunnart has been collected in south-western Papua New Guinea and on the western coast of Cape York Peninsula. It appears to be most closely related to the Carpentarian Dunnart and to have been confused with this species in the past: it is distinguishable from it by having large striated pads between the toes of its hindfeet. It differs from the Red-cheeked Dunnart, which also occurs in Cape York, in being smaller and lacking red cheeks. The Chestnut Dunnart has prominent dark rings around its eyes. Its generally chestnut colour is reflected in its common name.

Little is known of its biology but breeding appears to take place in the dry season between July and October. The female has six to eight teats and carries five to eight pouch-young.

HABITAT: tropical monsoonal savannah and woodland.

HEAD AND BODY: 8–11 cm

TAIL: 8–10 cm

DISTRIBUTION: 10,000–30,000 km^2

ABUNDANCE: very sparse

STATUS: vulnerable

Carpentarian Dunnart

Sminthopsis butleri (but'-ler-ee: "Butler's mouse-appearance", after H. Butler, Australian naturalist)

(R Whitford)

Almost nothing is known of the biology of this species. In captivity, it behaves like most other dunnarts, being a nocturnal predator on a wide variety of insects and other arthropods.

Mating probably occurs in spring.

HABITAT: tropical dry sclerophyll forest with dense grass understorey

HEAD AND BODY: about 9 cm

TAIL: about 9 cm

DISTRIBUTION: 30,000–100,000 km^2

ABUNDANCE: common

STATUS: probably secure

Fat-tailed Dunnart

Sminthopsis crassicaudata (kras'-i-kaw-dah'-tah: "fat-tailed mouse-appearance")

(R Whitford)

detach from the teats at four to five weeks and are suckled in a nest until nine to 10 weeks old.

HABITAT: cool-temperate to subtropical, arid to moderately wet woodland, shrubland and tussock grassland on clay, sand and stony desert
HEAD AND BODY: 6–9 cm
TAIL: 4–7 cm
DISTRIBUTION: more than 1 million km^2
ABUNDANCE: common
STATUS: secure

The common name of this species suggests that it is the only dunnart to store fat in the base of its tail, but this capacity is exhibited also by the White-tailed, Ooldea and Stripe-faced Dunnarts. It feeds at night on insects and arthropods and spends the day in a nest built under shelter or in cracks in the soil, often sharing it with several other individuals during the colder part of the year.

It does not need to drink.

Males are larger than females. Females become sexually mature at five to six months but possibly do not breed until more than 12 months old. Mating takes place from about May to about January and two litters are usually raised in a year. The female has eight to 10 teats in a well-developed pouch and the usual litter size is six to eight young. The young

Little Long-tailed Dunnart

Sminthopsis dolichura (dol'-ik-ue'-rah: "long-tailed mouse-appearance")

Nothing is known of the biology of this species.

HABITAT: temperate semiarid mallee
HEAD AND BODY: 7–8 cm
TAIL: 7–8 cm
DISTRIBUTION: 300,000–1 million km^2
ABUNDANCE: sparse
STATUS: probably secure

(DJ Kitchener)

Julia Creek Dunnart

Sminthopsis douglasi (dug'-las-ee: "Douglas's mouse-appearance", after A. Douglas, Australian naturalist)

This species is known from only four specimens. The only female among these had six young on seven teats.

HABITAT: probably subtropical woodland with dense grass cover
HEAD AND BODY: about 12 cm
TAIL: about 11 cm
DISTRIBUTION: 30,000–100,000 km²
ABUNDANCE: very rare
STATUS: possibly extinct

Gilbert's Dunnart

Sminthopsis gilberti (gil'-ber-tee: "Gilbert's mouse-appearance", after J. Gilbert, collector-naturalist for J. Gould)

Nothing is known of the biology of this species.

HABITAT: heathland, mallee, eucalypt woodland
HEAD AND BODY: 8–9 cm
TAIL: 7–8 cm
DISTRIBUTION: 30,000–100,000 km²
ABUNDANCE: common
STATUS: probably secure

(B & B Wells)

White-tailed Dunnart

Sminthopsis granulipes (gran'-ue-lee-pez: "granule-footed mouse-appearance")

(R Whitford)

One of the defining characters of this species is the even, rasp-like granulation of the soles of the fore and hind feet (and, hence, the absence of the usual pads). This is presumably an adaptation to the sandy substrate upon which the White-tailed Dunnart usually lives. It is a nocturnal predator on a wide variety of insects and other arthropods. There are indications that mating occurs in May and June.

HABITAT: cool-temperate low mallee shrubland, usually on sand
HEAD AND BODY: 7–9 cm

TAIL: 6–7 cm
DISTRIBUTION: 100,000–300,000 km²
ABUNDANCE: common
STATUS: probably secure

Grey-bellied Dunnart

Sminthopsis griseoventer (griz'-ay-oh-vent'-er: "grey-bellied mouse-appearance")

Nothing is known of the biology of this species.

HABITAT: warm-temperate to cool-temperate well-watered to semiarid eucalypt woodland, mallee, banksia scrub and heathland
HEAD AND BODY: 8–9 cm
TAIL: 8–9 cm
DISTRIBUTION: 30,000–100,000 km^2
ABUNDANCE: common
STATUS: probably secure

(DJ Kitchener)

Hairy-footed Dunnart

Sminthopsis hirtipes (her'-ti-pez: "hairy-footed mouse-appearance")

As indicated by its name, this species is characterised by having hairs between the granules of its soles and palms: a fringe of long hairs also extends outwards from the feet. This adaptation to locomotion on soft sandy surfaces is shared with the Lesser Hairy-footed Dunnart and the Sandhill Dunnart. Little is known of the biology of any of these species.

The female has six teats. Births probably take place in summer or autumn.

HABITAT: warm-temperate to cool-temperate, arid to semiarid woodland, shrubland and tussock grassland, usually on sand
HEAD AND BODY: 7–9 cm
TAIL: 7–10 cm
DISTRIBUTION: more than 1 million km^2
ABUNDANCE: sparse
STATUS: secure

(R Whitford)

White-footed Dunnart

Sminthopsis leucopus (lue'-koh-poos: "white-footed mouse-appearance")

This is the southernmost and most wet-adapted of the dunnarts, living in a variety of habitats which have dense undergrowth. It feeds on insects and other invertebrates but possibly also eats reptiles and the young of other small mammals. It is nocturnal, spending the day in a nest built under, or in, fallen timber or, rarely, in a tree.

Males are larger than females. Mating probably takes place from July to September. Females have 10 teats (mainland population) or eight (Tasmania). All teats may be occupied but the usual litter size is not known.

HABITAT: cool wet or dry sclerophyll forest to woodland, all with dense ground cover; coastal scrub, heathland, sedgeland and tussock grassland.
HEAD AND BODY: 7–12 cm
TAIL: 6–10 cm
DISTRIBUTION: 100,000–300,000 km²
ABUNDANCE: common
STATUS: secure

(R Whitford)

Long-tailed Dunnart

Sminthopsis longicaudata (lon'-jee-kaw-dah'-tah: "long-tailed mouse-appearance")

Almost nothing is known of the biology of this species, but its behaviour is different from that of other dunnarts. It leaps with agility among stones, balancing itself with a very mobile, tufted tail which is more than twice the length of the head and body. It is a nocturnal predator on insects or other invertebrates.

Males are slightly larger than females. Sexual maturity is reached at about eight to 11 months. Breeding probably occurs from August to December. The female has six teats in a well-developed pouch. Several litters may be reared successively.

HABITAT: subtropical stony lateritic breakaways and screes with woodland and hummock grassland
HEAD AND BODY: 8–10 cm
TAIL: 18–21 cm
DISTRIBUTION: less than 10,000 km²
ABUNDANCE: rare
STATUS: possibly endangered

(AG Wells)

Stripe-faced Dunnart
Sminthopsis macroura (mak'-roh-ue'-rah: "large-tailed mouse appearance")

In common with several other dunnarts, this species stores fat in the base of its long tail when food is abundant. It is a nocturnal predator on insects and other arthropods. During the day it shelters under stones or logs or in soil cracks.

Males are larger than females. Females reach sexual maturity at about five months. Mating takes place from July to February and two litters can be reared in a year. The female has eight teats and usually rears about six young, which detach from the teats at about six weeks and are suckled in a nest until about 10 weeks old.

HABITAT: tropical to warm-temperate arid to semiarid woodland, shrubland and tussock grassland
HEAD AND BODY: 7–10 cm
TAIL: 8–10 cm
DISTRIBUTION: more than 1 million km²
ABUNDANCE: sparse
STATUS: secure

(R Whitford)

Common Dunnart
Sminthopsis murina (myue-ree'-nah: "mouse-like mouse-appearance")

(R Whitford)

Mating extends from July to December and two litters may be reared in a year. The female has eight or 10 teats in a well-developed temporary pouch. Four to 10 young are carried in the pouch for about five weeks and subsequently suckled in a nest to the age of nine to 10 weeks.

HABITAT: warm-temperate to cool-temperate dry sclerophyll forest to woodland and heathland, mainly well-watered but extending to subarid
HEAD AND BODY: 6–10 cm
TAIL: 7–10 cm
DISTRIBUTION: more than 1 million km²
ABUNDANCE: common
STATUS: secure

The vernacular name of this species refers to its abundance in limited areas within its range and also to the fact that it is found near all the capital cities except Darwin and Hobart. Together with the White-footed Dunnart, with which it was once regarded as conspecific, it is characterised by living in relatively well-watered cool-temperate regions. Its patchy distribution may be due to its preference for areas that are in the process of regeneration after fire. It feeds at night on medium-sized insects and other invertebrates. During the day it sleeps in a nest constructed under shelter on the ground or in an excavation; sleep may be combined with torpor.

Males are slightly larger than females. It seems that sexual maturity is reached at about six months.

Ooldea Dunnart

Sminthopsis ooldea (ule'-day-ah: "Ooldea mouse-appearance". Ooldea is a town on the Nullarbor Plain in South Australia)

Little is known of the biology of this desert species. In laboratory conditions it readily kills and eats such relatively large insects as locusts. Well-nourished animals store fat in the base of the tail, which becomes swollen. It would not be surprising if the Ooldea Dunnart could survive without drinking water.

Males are slightly larger than females. Females become sexually mature at 10 months. Mating occurs from August to December, possibly with a peak in October. Females have eight teats in a temporary pouch and the usual litter is about seven young. These remain in the pouch for four to five weeks and are suckled in a nest until about 10 weeks old.

HABITAT: temperate arid woodland, shrubland, tussock grassland on sandy or stony soil
HEAD AND BODY: 5–9 cm
TAIL: 6–10 cm
DISTRIBUTION: 300,000–1 million km²
ABUNDANCE: common
STATUS: secure

(R Whitford)

Sandhill Dunnart

Sminthopsis psammophila (sam-off'-il-ah: "sand-loving mouse-appearance")

Little is known of the biology of this dunnart. It appears to be restricted to sand-dunes and, possibly, to be active by day. The hairs which cover the feet, including most of the sole of the hind foot, probably assist locomotion on fine sand. In captivity,

it has fed on insects and spiders. It is a large species, comparable in size to the Red-cheeked Dunnart.

One female specimen has eight teats. Nothing is known of the breeding.

HABITAT: low sand ridges with spinifex hummocks, separated by wide swales with sparse scrub, sometimes with clumps of mulga; cool-temperate to subtropical
HEAD AND BODY: 9–12 cm
TAIL: 11–13 cm
DISTRIBUTION: 10,000–30,000 km²
ABUNDANCE: very rare
STATUS: endangered

(R Ruehle)

Red-cheeked Dunnart

Sminthopsis virginiae (ver-jin'-ee-ee: "Virginia's mouse-appearance", significance unknown)

Little is known of the biology of this species. Like the Chestnut Dunnart it occurs in both Australia and New Guinea. Captive animals behave as typical dunnarts, being nocturnal predators on insects and other arthropods.

Sexual maturity is reached at about seven months. Captive animals have bred throughout most of the year. The female has eight teats. Several litters may be reared in succession.

HABITAT: tropical woodland, including swampy savanna
HEAD AND BODY: 8–13 cm
TAIL: 9–14 cm
DISTRIBUTION: 100,000–300,000 km²
ABUNDANCE: sparse
STATUS: probably secure

(R Whitford)

Lesser Hairy-footed Dunnart

Sminthopsis youngsoni (yung'-sun-ee: "Youngson's mouse-appearance", after W. K. Youngson of the Western Australian Museum)

This species, first described in 1982, is similar to but slightly smaller than the Hairy-footed Dunnart.

Mating certainly occurs in August and September but apparently not from April to June: there are no data for the other six months. Females have six teats and can carry six pouch-young.

HABITAT: arid tropical sand plains and dunes with tussock or hummock grasses
HEAD AND BODY: about 7 cm
TAIL: about 7 cm
DISTRIBUTION: 100,000–300,000 km²
ABUNDANCE: common
STATUS: probably secure

Family MYRMECOBIIDAE

(mer'-mek-oh-bee'-id-ee: "*Myrmecobius*-family")

The Numbat is the only member of this family, so the characteristics of the family are those of the species. It is the only marsupial to feed habitually upon social insects, a feature which it shares with the Short-beaked Echidna but, while the Echidna feeds mostly on ants, the Numbat's diet is almost exclusively termites.

Genus Myrmecobius

(mer'-mek-oh-bee'-us: "ant-living")

The characteristics of the genus are those of the species.

Numbat

Myrmecobius fasciatus (fas'-ee-ah'-tus: "striped feeder-on-ants")

The Numbat is the only marsupial that is specialised for feeding on small colonial insects. Like the Short-beaked Echidna, it picks these up with its long, rapidly moving, sticky tongue. Unlike the Echidna and eutherian anteaters, it does not have strongly developed digging fore-limbs. Its limbs are rather delicate, and the claws of its feet are not very large. It feeds mainly upon termites, which it finds in soft, rotting timber or in runways just below the surface of the ground, in which the termites travel between feeding areas and the nest.

Uniquely among the marsupials, it is habitually active by day. It sleeps at night in a hollow log or under fallen timber.

Sexual maturity is reached at about 11 months. The female has four teats but lacks a pouch. Mating occurs from December to February and three or four young are usually reared.

HABITAT: cool-temperate dry eucalypt forest and, in the nineteenth century, semiarid to arid woodland
HEAD AND BODY: 20–28 cm
TAIL: 16–21 cm
DISTRIBUTION: 10,000–30,000 km^2
ABUNDANCE: sparse
STATUS: endangered

(R Whitford)

Family THYLACINIDAE

(thie'-lah-sie'-nid-ee: "*Thylacinus*-family")

There is every indication that the member of this family that was described from Tasmania in the nineteenth century became extinct in the 1930s or 1940s. It was once widespread on the Australian mainland; it began to decline about five thousand years ago, but may have survived there until as recently as five hundred years ago.

Several fossil thylacinids have been described, including one from about 15 million years ago which differs only slightly from the Thylacine that recently became extinct. The characteristics of the family are essentially those of that species.

Genus *Thylacinus*

(thie'-lah-see'-nus: "pouched dog")

The most obvious characteristic of this genus is the overall wolf-like shape of the head and forequarters. The hindquarters are not wolf-like, and the rather inflexible tail has a very broad base.

Thylacine

Thylacinus cynocephalus (sie'-noh-sef'-al-us: "dog-headed pouched-dog")

The Thylacine is of great interest as the largest carnivorous marsupial to have existed in recent times and for its remarkable resemblance, particularly in the head and forequarters, to a wild dog. However, examination of the skull shows that the space occupied by the brain was considerably smaller than in a dog—a fact which may help to explain the demise of the Thylacine on mainland Australia after the arrival of the Dingo, probably about 5000 years ago. The Dingo did not reach Tasmania, which became separated from the mainland about 12 000 years ago.

We have little information on the biology of the Thylacine. It slept by day, probably in dense vegetation, and hunted other marsupials at night. It seems to have been unable to run fast but probably followed smaller animals until they became tired. When sheep were introduced into Tasmania, it attacked lambs and weak or penned sheep. The number killed seems not to have been large, but a substantial bounty was placed on Thylacine scalps and this led to heavy pressure on the species. However, it is unlikely that hunting and trapping were the main cause of its extermination. The rapid decline in the population is more likely to have been the result of alienation of its habitat for agriculture.

(J Wolf)

Breeding appears to have extended throughout the year, with a peak of births in winter and spring. Females had four teats in a backward-opening pouch. Up to four but usually two or three young were reared.

HABITAT: open forest and woodland with adjacent grassland
HEAD AND BODY: 100–130 cm
TAIL: 50–65 cm
DISTRIBUTION: nil
ABUNDANCE: nil
STATUS: extinct

Order PERAMELOMORPHIA

(pe'-rah-mel-oh-mor'-fee-ah: "*Perameles*-order")

In several respects the bandicoots and bilbies which comprise this order appear to be intermediate between the Dasyuromorphia and the Diprotodontia. They have four or five pairs of upper incisors and three pairs in the lower jaw but, unlike the condition in the Dasyuromorphia, these are stumpy rather than needle-like. As in the Diprotodontia, the second and third toes of the hind foot are fused together except for the claws. The fourth toe is longer and more powerfully muscled than the others and the fifth is small or absent. The forefoot is specialised for digging and bears strong, rather flattened claws on the second, third and fourth toes: the first and fifth toes are either small or absent.

In all peramelomorphs the snout is long and pointed and the neck is short. The head appears to continue into the body, which has large hindquarters, giving the animal a somewhat pear-shaped appearance. Food is obtained by digging with the forefeet. The diet varies between species but generally includes both small invertebrates and plant material.

All females have eight teats in a backward-opening pouch, but seldom rear more than four young. The embryo is connected to the maternal uterus by a placenta but, rather surprisingly, the gestation period is less than in those marsupials that lack a placenta. In fact, the Northern Brown Bandicoot and the Long-nosed Bandicoot have the shortest known gestation period for any mammal: twelve and a half days.

The Peramelomorphia contains only one superfamily, the Perameloidea.

Superfamily PERAMELOIDEA

(pe'-rah-mel-oy'-day-ah: "*Perameles*-superfamily")

The characteristics of this group are essentially those of the order Peramelomorphia. In the past, the superfamily was divided into two families: one for the typical bandicoots, the other for the bilbies, or rabbit-eared bandicoots. Here we adopt the recent view that a more natural division is into the primarily New Guinean spiny bandicoots (family Peroryctidae) and the remainder (family Peramelidae).

Family PERAMELIDAE

(pe'-rah-mel'-id-ee: "*Perameles*-family")

As interpreted here, the peramelids comprise the typical bandicoots, the bilbies and the Pig-footed Bandicoot (which probably became extinct early in the twentieth century). Typical bandicoots of the genera *Perameles* and *Isoodon* have short or "normal-sized" ears, short legs, stiff but rather sleek hair and a short, sparsely haired tail. Bilbies have very large ears, somewhat longer legs, silky fur and a rather long, well-furred tail. The Pig-footed Bandicoot had rather coarse hair, large ears, long legs and a rather long tail with a terminal brush. Typical bandicoots and bilbies are omnivorous; the Pig-footed Bandicoot was apparently herbivorous. In all peramelids, males are larger than females.

Genus *Chaeropus*

(kee'-roh-poos: "pig-foot")

The only known (now extinct) species of this genus had only three digits on the forefoot and four on the hind foot (all but the fourth being very reduced). The claws were rather hoof-like.

Pig-footed Bandicoot

Chaeropus ecaudatus (ay'-kaw-dah'-tus: "tail-less pig-foot", the first specimen to be described having accidentally lost its tail)

(*J Gould*)

twigs and grasses over a shallow scrape.

The female had eight teats in a backward-directed pouch but appears to have usually raised two young, born in May or June.

HABITAT: semiarid woodland with a dense understorey, shrubland and tussock grassland
HEAD AND BODY: 23–26 cm
TAIL: 10–15 cm
DISTRIBUTION: nil
ABUNDANCE: nil
STATUS: extinct

Unlike the other perameloids, the Pig-footed Bandicoot appears not only to have been quadrupedal but to have run on the tips of its hoof-like claws. Only the second and third toes of the forefeet were functional (the first and fifth being absent, the fourth minute). On the hind foot, only the fourth toe was functional (the first missing, the second and third fused and small, the fifth minute). Fragmentary evidence suggests that this species was also unusual in eating grasses, but captive animals accepted grasshoppers. By day it slept in a typical bandicoot nest of

Genus *Isoodon*

(ie-soh'-oh-don: "equal-tooth")

Members of this genus differ from species of *Perameles* in having somewhat shorter, less acutely pointed snouts and relatively shorter and more rounded ears. The name refers to the similar size of the incisor teeth, a feature of all bandicoots. The hind foot has five toes.

Golden Bandicoot

Isoodon auratus (or-ah'-tus: "golden equal-tooth")

This shiny russet-brown bandicoot is the smallest member of the genus *Isoodon*. It appears to forage in the usual bandicoot manner and is known to eat a wide variety of invertebrates including beetle larvae, termites, moths and centipedes as well as succulent tubers. It sleeps by day in a simple nest of grass, in a cave or under a tussock.

Breeding extends throughout the year. The female has eight teats but normally rears only two young.

There has been a great reduction in distribution since the 1930s and the species is now restricted to a relatively small area of the coastal Kimberley region and Barrow Island. The island form is distinctly smaller than that of the mainland.

HABITAT: tropical woodland to spinifex grassland
HEAD AND BODY: 19–30 cm
TAIL: 8–12 cm
DISTRIBUTION: 30,000–100,000 km²
ABUNDANCE: sparse
STATUS: vulnerable

(K Johnson)

Northern Brown Bandicoot

Isoodon macrourus (mak'-roh-yue'-rus: "long-tailed equal-tooth")

The Northern Brown Bandicoot closely resembles its southern relative but is about twice its weight (males may exceed 2 kilograms). It forages at night in the usual bandicoot manner for subterranean insects, spiders and earthworms and tubers; it also eats some berries and seeds. By day it sleeps in a nest of sticks, vegetation and earth, usually with a defined exit and entry. Both sexes are solitary, territorial and aggressive.

In the tropical and subtropical parts of the range, breeding is continuous throughout the year. At the southern extreme, there is little or no breeding in autumn. The female has eight teats but usually rears about four young. A female can produce her first litter at the age of about four months and thereafter produce litters at intervals of about seven weeks.

A supposed species, *I. arnhemensis*, was described in 1981 from Melville Bay, NT. It is not considered to be sufficiently well-founded to be separable from *I. macrourus*.

(AC Robinson)

HABITAT: tropical to subtropical wet and dry sclerophyll forest to woodland, with dense grass or shrub ground cover
HEAD AND BODY: 30–47 cm
TAIL: 8–22 cm
DISTRIBUTION: 300,000–1 million km²
ABUNDANCE: common
STATUS: secure

Southern Brown Bandicoot

Isoodon obesulus (ob-es-ue'-lus: "somewhat-fat equal-tooth")

This powerfully built bandicoot forages at night for subterranean insects and earthworms by digging conical pits with its forepaws. During the day it sleeps under a well-constructed nest of grass, twigs and earth which, under wet conditions, may be built on an elevated platform of soil. Individuals of both sexes are solitary, territorial and aggressive.

Breeding extends from about May to September. A female has eight teats but usually rears two to four young. Sexual maturity is reached at the age of three to four months and a female may rear three litters a year.

HABITAT: wet and dry sclerophyll forest to woodland, usually with understorey of rather dense scrub which is periodically burned

(HJ Aslin)

HEAD AND BODY: 28–36 cm
TAIL: 9–14 cm
DISTRIBUTION: 300,000–1 million km²
ABUNDANCE: sparse
STATUS: secure

Genus Macrotis

(mak-roh'-tis: "big-ear")

Bilbies have the same general head and body shape as typical bandicoots but their ears are long and rather rabbit-like. Their size may well assist the sense of hearing but they also probably act as radiators of body heat during the day. The long silky fur probably acts as a thermal insulation, which may well be valuable in the cold desert nights when they forage for food. Unlike typical bandicoots, bilbies have a rather long and mobile tail with a bushy tip. They have longer limbs than typical bandicoots and move with a more quadrupedal gait (which nevertheless looks rather clumsy). The hind foot has four toes. Apart from wombats, bilbies are the only marsupials to dig substantial burrows and, like wombats, they shelter in these during the day.

Bilby

Macrotis lagotis (lag-oh'-tis: "hare-eared big-ear")

With its long ears and silky fur, the Bilby cannot be confused with any other Australian mammal (except the Lesser Bilby, which appears to be extinct). It forages at night in the usual bandicoot manner for subterranean insects, bulbs and fungi but it also eats some fruits and (unusually for a perameloid) seeds. The feeding excavations that it makes with its forefeet are deeper than those of typical bandicoots. It obtains sufficient water from its food to make drinking unnecessary. By day it sleeps in a deep burrow, the entrance of which is usually beside a tussock, shrub or termite mound.

Males are much larger than females. Breeding extends throughout the year, but may be dependent upon an adequate food supply. The female has eight teats in a backward-opening pouch but normally rears two young.

HABITAT: arid to semiarid woodland, shrubland and hummock grassland, particularly areas that are regenerating after fire
HEAD AND BODY: 30–55 cm
TAIL: 20–30 cm
DISTRIBUTION: 300,000–1 million km²
ABUNDANCE: very sparse
STATUS: endangered

(K Johnson)

Lesser Bilby

Macrotis leucura (luke-ue'-rah: "white-tailed big-ear")

Similar in body shape to the Bilby, the Lesser Bilby is a much smaller species, about one-third its weight. It appears to have become extinct in the 1930s. Fragmentary evidence indicates that it was largely carnivorous, feeding on rodents and also eating seeds and vegetation.

Like the Bilby, it constructed a deep burrow, but the entrance was sealed with sand when the animal was inside during the day.

Females had eight teats and are reported to have reared two young at a time.

HABITAT: desert sandhills
HEAD AND BODY: 20–27 cm
TAIL: 12–17 cm
DISTRIBUTION: nil
ABUNDANCE: nil
STATUS: extinct

(O Thomas)

Genus Perameles

(pe'-rah-mel'-ayz: "pouched-badger")

Members of the genus *Perameles* are typical bandicoots, characterised by having a somewhat larger and more acutely pointed snout than members of the genus *Isoodon*.

The rear half of the sole of the hind foot is covered with hairs. The hind foot has five digits.

Western Barred Bandicoot

Perameles bougainville (bue'-gan-veel: "Bougainville's pouched-badger", after Baron Bougainville, French navigator)

(AC Robinson)

The Western Barred Bandicoot is more delicately built than its eastern relative and has proportionally larger ears. It has two or three transverse bars of paler fur across its rump but these are less well defined than in the eastern species. It is relatively small, less than a quarter of the weight of the Long-nosed Bandicoot. At night, it forages for subterranean insects, other invertebrates and succulent roots, also eating some green herbage. During the day it sleeps in a nest of vegetation over a shallow burrow, usually in the shade of a shrub. Once widespread over southern and western Australia, it is now restricted to relict populations on Bernier and Dorre Islands in Shark Bay.

Breeding occurs in autumn and winter. The female has four or eight teats but usually rears two or three young.

HABITAT: semiarid woodland, shrubland and dunes
HEAD AND BODY: 20–30 cm
TAIL: 8–12 cm
DISTRIBUTION: 10,000–30,000 km^2
ABUNDANCE: abundant
STATUS: vulnerable

Desert Bandicoot

Perameles eremiana (e'-rem-ee-ah'-nah: "desert pouched-badger")

Not collected since the 1930s, this small bandicoot (a little larger than the Western Barred species) had rather long ears and hairy soles to the fore and hind feet. It slept by day in a nest of vegetation made in a scrape in the sand. It had eight teats and is reported to have reared two young in a litter.

HABITAT: arid spinifex grassland
HEAD AND BODY: 24–28 cm
TAIL: 12–14 cm
DISTRIBUTION: nil
ABUNDANCE: nil
STATUS: extinct

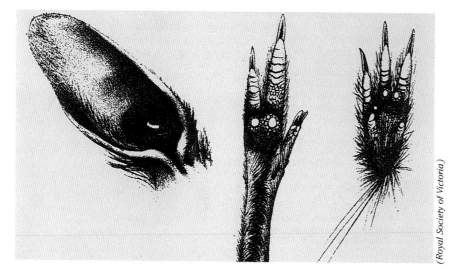

(Royal Society of Victoria)

Eastern Barred Bandicoot

Perameles gunnii (gun'-ee-ee: "Gunn's pouched-badger", after R. Gunn, who collected the first specimen)

(JE Wapstra)

rears two or three young. Under favourable conditions, a female begins to breed at the age of about three months and thereafter produces a litter every seven weeks. Males become sexually mature at four to five months.

HABITAT: seasonally wet woodland and grassland
HEAD AND BODY: 27–37 cm
TAIL: 7–11 cm
DISTRIBUTION: 30,000–100,000 km²
ABUNDANCE: common
STATUS: probably secure

A little smaller than the Long-nosed Bandicoot, the Eastern Barred Bandicoot has much the same body shape but is a more attractive animal with four pale transverse bars on its rump. It forages at night for burrowing insects, worms and succulent roots and corms by digging conical pits in the soil with its long-clawed forefeet. It is also known to eat berries. By day, it sleeps under a loose pile of grass over a shallow scrape in the ground. Both sexes are solitary, territorial and aggressive.

In Victoria, breeding takes place throughout the year but is greatest from midwinter to early summer: in Tasmania, breeding occurs in winter, spring and summer but not autumn. The female has eight teats but usually

Long-nosed Bandicoot

Perameles nasuta (naz-ue'-tah: "prominent-nosed pouched-badger")

The Long-nosed Bandicoot is a compact animal with moderate-sized ears. It forages at night for burrowing insects, other soil-dwelling invertebrates and succulent corms or roots, by digging conical holes with its forepaws. During the day it sleeps in a rough nest of vegetation over a shallow scrape in the soil. Both sexes are solitary and aggressive.

Breeding takes place throughout the year, but to a lesser extent in winter. The female has eight teats but usually rears two or three young. Under favourable conditions, a female produces her first young at the age of five months and thereafter, in favourable conditions, produces a litter every seven weeks.

(R Miller)

HABITAT: rainforest, wet and dry sclerophyll forest, well-watered woodland
HEAD AND BODY: 31–43 cm
TAIL: 12–16 cm
DISTRIBUTION: 300,000–1 million km²
ABUNDANCE: common
STATUS: secure

Family PERORYCTIDAE

(pe'-roh-rik'-tid-ee: "*Peroryctes*-family")

Four species of spiny bandicoots of the genus *Peroryctes* are found in New Guinea. A related genus, *Echymipera*, has three species in New Guinea and one of these also occurs on Cape York Peninsula.

Peroryctids differ from peramelids in having only four pairs of upper incisors (peramelids have five pairs) and in having a very short tail and spiny hair. Other differences in blood chemistry that have recently been demonstrated suggest that, despite their superficial similarity to typical bandicoots, the members of this essentially New Guinean group are more different from the peramelids than the various peramelids are from each other.

Genus Echymipera

(ek'-im-i-pe'-rah: "pouched-hedgehog")

Members of this genus are large (up to 2 kilograms) and are characterised by rather spiny, glossy hair. They have only four pairs of upper incisors.

Rufous Spiny Bandicoot

Echymipera rufescens (rue-fes'-enz: "reddish pouched-hedgehog")

The Rufous Spiny Bandicoot forages at night among the litter on the forest floor for insects and other invertebrates. Little is known of its biology.

The female has eight teats in a backward-opening pouch.

HABITAT: lowland tropical rainforest to woodland and adjacent heath
HEAD AND BODY: 30–40 cm
TAIL: 8–10 cm
DISTRIBUTION: 10,000–30,000 km²
ABUNDANCE: common
STATUS: probably secure

(GC Richards)

Order DIPROTODONTIA

(die-proh'-toh-don'-tee-ah: "two-front-toothed", as in the fossil *Diprotodon*)

Members of this group are restricted to Australia, New Guinea, Indonesia eastward of Bali and Melanesia westward of the Solomons. Diprotodonts never have more than one *functional* pair of incisors in the lower jaw, but vestigial second incisors are sometimes present; there are usually three upper pairs of incisors, but wombats have only one pair. When present, the upper canines are usually small; lower canines are absent. Diprotodonts are primarily herbivorous, but some species are insectivorous, some eat insects and other small animals, while some are omnivorous. A few feed on pollen and nectar and the Honey-possum eats nothing else.

The second and third toes of the hind foot are fused together to form what appears to be one digit with two claws (a condition known as syndactyly).

The living diprotodonts are classified into two suborders: the Vombatiformes, comprising the wombats, koala and various fossil groups, including the giant diprotodontids and palorchestids; and the Phalangerida, comprising all of the possums and kangaroo-like marsupials.

Suborder VOMBATIFORMES

(vom'-bah-tee-for'-mays: "*Vombatus*-suborder")

Of the seven families in this relic group, five are extinct, the only two surviving being the Vombatidae, with three living species; and the Phascolarctidae, with only one. While the wombats and Koala are more closely related to each other than to any other living marsupials, current opinion is that the difference between them is such that they should be placed in separate infraorders, Vombatomorphia and Phascolarctomorphia.

Infraorder PHASCOLARCTOMORPHIA

(fas'-koh-lark'-toh-mor'-fee-ah: "*Phascolarctos*-infraorder")

This group includes only one living species and up to a dozen fossil species, extending over the past fifteen million years. Since most of these fossils consist only of teeth and jaw fragments, we know nothing of the body of these older forms, but the teeth demonstrate that they all browsed on leaves and that all were much the same size as the modern Koala—the oldest being rather smaller, the most recent being larger.

Family PHASCOLARCTIDAE

(fas'-koh-lark'-tid-ee: "*Phascolarctos*-family")

This family has only one living species, the Koala. Like wombats, the Koala has stout, rodent-like incisors, but three upper pairs work against one lower pair. It feeds on the leaves of trees, which are less abrasive than grasses, and its teeth do not grow continuously like those of wombats.

The Koala has large, strongly clawed fore and hind feet. The forefoot has a very marked division between the first two digits and the other three—a "two-thumbed" or "split-hand" arrangement which enables it to grip a branch in a manner similar to a parrot. The hind foot has a large and powerful thumb-like first digit which can grip against the sole and the other toes. It is thus a very sure-footed climber which manages very well in the absence of a prehensile tail. Even without utilising its grip, the Koala can climb the trunk of a tree by digging in its claws.

Although the Koala has been referred to as a "Marsupial sloth", it can be extremely active, often leaping from one tree to another. It would be more appropriate to compare it with the Indri, an agile Madagascan lemur.

Genus Phascolarctos

(fas'-koh-lark'-tos: "pouched-bear")

The characteristics of the genus are essentially those of the species.

Koala

Phascolarctos cinereus (sin-er-ay'-us: "ash-coloured pouched-bear")

Because the Koala feeds almost exclusively on the leaves of eucalypts and exercises considerable choice among these, it is restricted to limited areas of forest. It may show some activity by day but it usually spends the daylight hours asleep in the fork of a tree: it does not make a nest or seek any shelter.

It moves around vigorously at night, travelling to favoured food trees and often acting aggressively towards other individuals. In open forest or woodland, it may descend to the ground and even swim across watercourses to move from its base tree to feeding trees. Males are somewhat larger than females and are more aggressive. Older males attempt to monopolise several females.

Koalas from the southern part of the extensive distribution are much larger (almost twice the average weight) than those from the most northern part. Males are larger than females. Sexual maturity is reached at two years of age but males seldom have the opportunity to mate before the age of three or four. Most mating occurs from October to February. The female has two teats in a pouch which opens downwards and to the rear. Only one young is born and after five or six months in the pouch it is carried on the mother's back.

HABITAT: cool-temperate to tropical wet and dry eucalypt forests and woodland
HEAD AND BODY: 50–82 cm
TAIL: negligible
DISTRIBUTION: 300,000–1 million km^2
ABUNDANCE: very sparse
STATUS: probably secure

(*LF Schick*)

Infraorder VOMBATOMORPHIA

(vom'-bah-toh-mor'-fee-ah: "*Vombatus*-infraorder")

This group comprises five fossil families (including the giant diprotodons and the marsupial lions), and one living family, the Vombatidae (wombats), all relatively large to immense marsupials which, with the exception of the marsupial lions (*Thylacoleo*), appear to have been herbivorous. The living wombats, and a number of extinct species, comprise the family Vombatidae.

Family VOMBATIDAE

(vom-bah'-tid-ee: "*Vombatus*-family")

Wombats have a short broad head which accommodates the powerful muscles needed to cut and grind coarse vegetation. The incisors have been reduced to one very large pair in the upper jaw and a matching pair below (an arrangement like that of rodents). There are no canine teeth. The incisors and the grinding teeth have open roots and grow continuously as they are worn down (also like those of rodents). The limbs are short and powerful and armed with broad claws. The first digit of the hind foot is reduced to a stub. The pouch of the female opens to the rear.

Wombats dig deep and often complex burrows in which they shelter during the day.

There are only two genera of living wombats: *Vombatus* (with one species) is a forest-dweller which grazes in clearings or at the forest edge; *Lasiorhinus* (with two species) inhabits more arid woodland and has a more varied diet.

Genus Lasiorhinus

(la'-zee-oh-rie'-nus: "hairy-nose")

Members of this genus differ from *Vombatus* in having fine silky fur and a hairy muzzle. They inhabit semiarid woodland rather than forests.

Northern Hairy-nosed Wombat

Lasiorhinus krefftii (kref'-tee-ee: "Krefft's hairy-nose", after G. Krefft, who forwarded a skull of the species to the describer, R. Owen)

In the nineteenth century this species was present in New South Wales and Victoria. It now survives only in a small national park near Epping Forest Station in tropical Queensland. At night it feeds on coarse grasses and herbs. By day it sleeps in a burrow.

Its reproductive biology is not known.

HABITAT: subtropical semiarid woodland and grassland on sandy soil
HEAD AND BODY: 80–100 cm
TAIL: 3–5 cm
DISTRIBUTION: less than 10,000 km²
ABUNDANCE: very sparse
STATUS: endangered

(CA Henley)

Southern Hairy-nosed Wombat

Lasiorhinus latifrons (lat'-i-fronz: "broad-headed hairy-nose")

The Southern Hairy-nosed Wombat lives mainly on the Nullarbor Plain. It has an effective water economy and does not need to drink. At night it feeds on coarse grasses and herbs and by day it rests well below the desert surface in a humid burrow.

Survival in this very difficult environment is assisted by a very low rate of metabolism.

Sexual maturity is reached at the age of three years. Most mating takes place from August to November but reproduction ceases in periods of low rainfall, when food is scarce. The rear-opening pouch has two teats, but usually one young is born. This remains in the pouch for up to nine months.

HABITAT: temperate semiarid woodland and shrubland on rather friable soil
HEAD AND BODY: 77–100 cm
TAIL: 2–6 cm
DISTRIBUTION: 30,000–100,000 km²
ABUNDANCE: sparse
STATUS: vulnerable

(B & B Wells)

Genus *Vombatus*
(vom-bah'-tus "wombat")

The single species of this genus differs from the two species of *Lasiorhinus* in having a naked muzzle and rather coarse body hair.

Common Wombat
Vombatus ursinus (er-see'-nus: "bear-like wombat")

The Common Wombat has coarse, stiff hair and a naked muzzle. During the night it moves over a large area, grazing on young grasses and sedges. During the day it sleeps in a nest of vegetation in one of several burrows (constructed by itself or other wombats) within its home range. In cold weather it may feed during the day. Both sexes are solitary and aggressive but their activities are organised so that confrontations are infrequent.

Sexual maturity is reached at the age of two years. Breeding takes place throughout the year but with a peak in September, October and November. The female has two teats in a backward-opening pouch but usually rears only one young, which remains in the pouch for about six months, after which it is able to follow its mother, on foot, until it is weaned and independent.

HABITAT: wet and dry sclerophyll forest with adjacent grassy areas
HEAD AND BODY: 90–120 cm
TAIL: 2–3 cm
DISTRIBUTION: 100,000–300,000 km²
ABUNDANCE: sparse
STATUS: secure

(D Watts)

Suborder PHALANGERIDA
(fal'-an-je'-rid-ah: "*Phalanger*-suborder")

This recently erected group includes the possums, gliders, cuscuses and kangaroo-like marsupials—all those diprotodont marsupials that are not members of the Vombatiformes. Although well supported by immunological evidence, the Phalangerida evades rigid definition on meaningful anatomical grounds and the concept leaves many questions of relationships unanswered. Nevertheless, the Phalangerida is not a rag-bag or "group of convenience". On the contrary, it draws attention to the invalidity of long-standing views: (a) that the "possums" comprise a closely-related group; and (b) that these are notably distinct from the Kangaroos. The new classification implies that what we call "possums" (including cuscuses and gliders) comprise a number of quite distinct groups and that the relationship between some of these may be no closer than between some "possums" and the more primitive kangaroos. Just what these relationships are remains to be discovered.

On present evidence, the living members of the Phalangerida are divided into five superfamilies. The Phalangeroidea comprises a single family, the Phalangeridae (cuscuses and brushtail possums). The Burramyoidea has only one family, the Burramyidae (pygmy-possums). The Petauroidea includes two families: the Petauridae (most gliders and Leadbeater's Possum); and the Pseudocheiridae, (ringtail possums and the Greater Glider). The Tarsipedoidea includes two families: the Tarsipedidae (Honey-Possum) and the Acrobatidae (feathertails). The Macropodoidea includes two families: the Potoroidae (rat-kangaroos, potoroos and bettongs); and the Macropodidae (wallabies and kangaroos).

Superfamily PHALANGEROIDEA

(fal'-an-je-roy'-day-ah: "*Phalanger*-superfamily")

As used here, the Phalangeroidea is restricted, among the living marsupials, to the cuscuses and brushtail possums. Inasmuch as the Phalangeroidea was once regarded as including all the possums, this is a little confusing, but the situation has arisen from increasing recognition that, in its older sense, the Phalangeroidea was not a natural grouping. With the separation of the "possums" into four superfamilies, the Phalangeroidea has shrunk to embrace one family, the Phalangeridae.

Family PHALANGERIDAE

(fal'-an-je'-rid-ee: "*Phalanger*-family")

The cuscuses, Scaly-tailed Possum and brushtail possums have similar dentition and skull structure, but they vary considerably in other anatomical features. Most cuscuses have a rather flat face with large, forward-directed eyes, a strongly prehensile tail, and forefeet that can oppose the first two fingers against the other three to grip a branch. The Scaly-tailed Possum has a pointed snout, large forward-directed eyes, forefeet like those of a cuscus and a strongly prehensile tail. Brushtails have a longer snout (described as "fox-like") and somewhat more laterally directed eyes; the tail is only weakly prehensile and cannot support the weight of the body; and the fingers of the hand grip by moving inwards towards the palm. All female phalangerids have a well-developed, forward-opening pouch which encloses two or four teats.

Genus Phalanger

(fal'-an-jer: "notable-digits", referring to fused second and third toes of hind foot)

Most cuscuses live in New Guinea and adjacent islands, extending westward to Sulawesi. The two species on Cape York Peninsula are outliers of populations that are widespread in New Guinea. Cuscuses inhabit tropical forest (mainly rainforest) canopy, feeding on fruits, flowers and leaves, supplemented by large insects, eggs and nestling birds. They move deliberately, gripping with forefeet in which the first two toes can be opposed to the other three, and hind feet with a large opposable first toe. All toes except this last are armed with sharp, curved claws. The tail is strongly prehensile and up to half of its distal end is bare and scaly, with horny projections on the underside.

The thick and woolly fur is strikingly patterned in patches of dark brown, orange and white. The metabolic rate is lower than in most other marsupials. Some cuscuses sleep in tree hollows, others do not utilise any shelter.

Spotted Cuscus

Phalanger maculatus (mak'-yue-lah'-tus: "spotted notable-digit")

(G. Schick)

This cuscus is widespread in New Guinea but, in Australia, is limited to the tip of Cape York Peninsula. Animals from New Guinea, particularly males, may be strikingly patterned in orange and white but those in Australia tend to be greyish with white patches above and white below. The face is round and the ears are barely visible.

The Spotted Cuscus is known to eat fruits and leaves and, in captivity, is partial to meat and eggs; it is therefore likely that it eats eggs, nestling birds and other small animals. It is nocturnal but does not utilise any shelter during the day. The female has four teats: up to three-pouch-young may be carried but usually only one young is reared. After leaving the pouch, it is carried on the mother's back. Breeding appears to take place throughout the year.

HABITAT: lowland tropical rainforest and adjacent mangroves
HEAD AND BODY: 35–45 cm
TAIL: 32–43 cm
DISTRIBUTION: 10,000–30,000 km^2
ABUNDANCE: very sparse
STATUS: vulnerable

Grey Cuscus

Phalanger orientalis (o'-ree-en-tah'-lis: "eastern notable-digit")

Most widespread of all the cuscuses, this species extends from the Moluccas through New Guinea to the Solomons. The Australian population lives at the base of Cape York Peninsula. The Grey Cuscus is smaller than the Spotted Cuscus, and has larger ears and a longer snout. It is greyish brown with a distinct dark brown stripe from between the eyes to the base of the tail.

It feeds on leaves and fruits and will eat meat in captivity, but not to the same extent as the Spotted Cuscus. It is nocturnal and seeks shelter during the day in a tree-hole. The female has four teats and may carry up to three pouch-young, although only one is usually reared. After leaving the pouch, it is carried on its mother's back until weaned. The Grey Cuscus probably breeds throughout the year.

HABITAT: tropical rainforest

HEAD AND BODY: 35–40 cm

TAIL: 28–35 cm

DISTRIBUTION: 10,000–30,000 km²

ABUNDANCE: very sparse

STATUS: vulnerable

(Queensland National Parks and Wildlife Service)

Genus *Trichosurus*

(trik'-oh-sue'-rus: "hairy-tail")

Together, the three species of brushtail possums occupy almost all parts of Australia where there are trees. They are adaptable herbivores, feeding largely on leaves but also eating fruits, buds, lichens and fungi. Although their tails are only weakly prehensile, they are active climbers and often leap from one branch to another, crashing into the foliage. Unlike cuscuses and the Scaly-tailed Possum, they do not have a "split hand", all fingers instead gripping towards the palm; the hind foot has a large and powerful opposable first toe.

Brushtails are very catholic in their choice of sites in which to sleep during the day. Tree-holes are used when available, but otherwise they may sleep in hollow logs, in crevices between roots or under rocks. The Common Brushtail often utilises spaces under the roofs of buildings.

Northern Brushtail Possum

Trichosurus arnhemensis (arn'-em-en-sis: "Arnhem [Land] hairy-tail")

Closely related to the Common Brushtail, this species occupies the north-western part of tropical Australia. It feeds mainly upon leaves, but also eats fruits and flowers. It is an opportunistic herbivore, feeding on whatever leaves are most readily available: it is quite tolerant of a number of toxic substances found in leaves.

In appearance, it closely resembles the Common Brushtail but it is noticeably smaller and its tail is less furred, particularly on the underside. The grey fur has a russet tinge. Its general biology appears to be very similar to that of the Common Brushtail but it may be more dependent upon tree-holes for nesting, and it is less aggressively territorial.

Breeding takes place throughout the year. The female has two teats but usually raises only one young, which is carried on the mother's back after it leaves the pouch.

HABITAT: tropical dry sclerophyll forest and woodland
HEAD AND BODY: 36–46 cm
TAIL: 24–30 cm
DISTRIBUTION: more than 1 million km²
ABUNDANCE: sparse
STATUS: secure

(JA Kerle)

Mountain Brushtail Possum

Trichosurus caninus (kah-nee'-nus: "dog-like hairy-tail")

Largest of the brushtails, this species is largely confined to south-eastern Australia on the coastal side of the Great Dividing Range. Despite its common name, it is not restricted to high country, but most of the coastal forests which once provided a habitat have been cleared and it is therefore more common in the highlands. It can be distinguished from the Common Brushtail by its plumper body and more bushy tail.

It feeds mainly upon leaves of shrubs, particularly acacias, supplemented by fruits and fungi; it is sometimes a pest of pine plantations. During the day it rests in tree-holes or hollow logs. The breeding season is from March to May. The female has two teats but usually raises only one young, which is carried on the mother's back after it leaves the pouch.

HABITAT: cool rainforest and wet sclerophyll forest
HEAD AND BODY: 40–50 cm
TAIL: 34–42 cm
DISTRIBUTION: 300,000–1 million km²
ABUNDANCE: common
STATUS: secure

(GA Hoye)

Common Brushtail Possum

Trichosurus vulpecula (vool-pek'-ue-lah: "little-fox-like hairy-tail")

(*G Weber*)

its territory with secretions from glands on its chest and rump and vigorously defending it against intruders, first with hisses and barks and then, if necessary, with tooth and claw.

The female has two teats but carries only one pouch-young. Breeding occurs throughout the year but with peaks in spring and autumn.

Introduced into New Zealand in 1937, the Common Brushtail has there become a serious pest.

HABITAT: wet and dry sclerophyll forest and woodland
HEAD AND BODY: 35–55 cm
TAIL: 25–40 cm
DISTRIBUTION: more than 1 million km²
ABUNDANCE: abundant
STATUS: secure

Because the range of this species includes every Australian capital city except Darwin (where its place is taken by the closely related Northern Brushtail), and because of its tendency to make its den in the roof-spaces of houses, it is the most familiar of all Australian marsupials. In suburbs it may eat introduced fruits and flowering shrubs but, in the wild, it feeds extensively on leaves, supplemented by fruits, buds and grasses; the diet varies considerably, depending upon the dominant vegetation. It prefers to make a nest in a tree-hole but, in the absence of trees of sufficient size or age, it will make a nest under any suitable shelter on the ground.

The Common Brushtail varies considerably in size, being much larger in the southern part of its range than in the north. A subspecies in northern Queensland has coppery red fur.

It is solitary and territorial, marking

Genus *Wyulda*

(wie-ool'-dah: "brushtail possum")

This genus is very different from the brushtail possums, but owing to a misunderstanding it was given an Aboriginal name for the brushtail.

The characteristics of the genus are those of its single species. It is distinguished by the rasp-like scales which cover most of its tail.

Scaly-tailed Possum

Wyulda squamicaudata (skwah'-mee-kaw-dah'-tah: "scaly-tailed brushtail")

Little is known of the biology of this species, which was not described until 1917 and was not observed in the wild until relatively recently. As the name implies, most of the strongly prehensile tail is covered with scales, each of which has a conical projection, giving the tail a rasp-like appearance. As in cuscuses, the first two fingers of the forefoot can be opposed to the other three and the hind foot has a large, opposable first toe. It is an agile climber and feeds at night on leaves, fruits and flowers. During the day it sleeps on the ground in rock crevices or among piled boulders.

The female has two teats but carries only one pouch-young.

HABITAT: tropical open woodland and vine-thickets in rocky country
HEAD AND BODY: about 40 cm
TAIL: about 30 cm
DISTRIBUTION: 30,000–100,000 km^2
ABUNDANCE: rare
STATUS: vulnerable

(E Beaton)

Superfamily BURRAMYOIDEA

(bu'-rah-mie-oy'-day-ah: "*Burramys*-superfamily")

The characteristics of the superfamily are those of the family. It may be noted, however, that most classifications include the Feathertail Glider (and the non-gliding Feathertail Possum of New Guinea) within the group. These two species are now placed in a family of their own, the Acrobatidae.

Family BURRAMYIDAE

(bu'-rah-mie'-id-ee: "*Burramys*-family")

This includes the terrestrial Mountain Pygmy-possum (*Burramys*) and four arboreal pygmy-possums of the genus *Cercartetus*, one of which also occurs in New Guinea. As the common names imply, these are very small marsupials (10 to 50 grams). The strongly prehensile tail is longer than the mouse-like head and body—about 25 per cent longer in the Long-tailed Pygmy-possum. The molar teeth have low smooth cusps, but one of the upper premolars on each side is a serrated, squarish blade-like structure that can be used to cut tough food into small fragments. A premolar of similar structure is found in the more primitive members of the kangaroo superfamily.

As a group, burramyids occupy a wide range of habitats from tropical rainforest to semiarid heathland and alpine shrublands. They are generally insectivorous, but the Eastern Pygmy-possum eats a great deal of nectar and pollen and the Mountain Pygmy-possum includes seeds in its diet.

It is a general characteristic of marsupials that the smaller species tend to have more teats and larger litters than the bigger ones. Smaller species also tend to have less well-developed pouches. This is demonstrated in the pygmy-possums, which may have as many as six teats and where the pouch is weakly developed in the smallest species, *Cercartetus lepidus*.

All pygmy-possums make a nest of shredded bark and leaves, usually situated in a tree-hole. All become torpid for short periods when food is scarce or the temperature is low.

Genus *Burramys*

(bu'-rah-mis: "stony-place-mouse", burra-burra means "stony place" in the language of the Aborigines living near the Wellington Caves, NSW)

Mouse-like in appearance, *Burramys* has a long tail and is most readily recognised by its very large blade-like premolar teeth. In size, shape and grooving, these are very similar to the cutting premolars of the Musky Rat-kangaroo and other potoroids.

Mountain Pygmy-possum

Burramys parvus (par'-vus: "small stony-place-mouse")

Largest of the pygmy-possums, this species is currently restricted to small areas above the snowline in the vicinity of Mount Hotham, Mount Higginbotham and Mount Kosciusko in the Australian Alps, but it had a much wider distribution in eastern Australia some 50 000 years ago and was first described, in the nineteenth century, from Pleistocene fossils in Wellington Caves.

It builds a more or less spherical nest of dry grass. During the warmer parts of the year, it forages on the ground and in the vegetation, feeding mainly upon insects, spiders and worms. Seeds are gathered during this period of abundance and cached under the nest or under nearby bark or stones. When snow lies a metre or more deep over the shrubs, the Mountain Pygmy-possum continues to be active in runways close to the ground and ekes out its meagre food supply by metabolising its stored fat and eating seeds which it stored in the summer. It may also become torpid for several days at a time, permitting its body temperature to drop almost to that of the surrounding air, and thereby conserving energy.

Males are slightly larger than females. Mating appears to take place in October and November. The female has four teats and usually rears four young, which remain in the nest for about a month after vacating the pouch.

HABITAT: alpine boulder moraines with dense heath cover and usually low woodlands
HEAD AND BODY: 10–12 cm
TAIL: 14–15 cm
DISTRIBUTION: less than 10,000 km²
ABUNDANCE: sparse
STATUS: possibly endangered

(R Miller)

Genus *Cercartetus*

(ser'-kar-tay'-tus: meaning unclear, possibly intended as cercaërtus, "tail-in-air")

Members of this genus are small, with rather a mouse-like head and body and a long, prehensile tail. They feed on insects, flowers, nectar and pollen.

Long-tailed Pygmy-possum

Cercartetus caudatus (kaw-dah'-tus: "tailed *Cercartetus*")

Largest member of the genus *Cercartetus*, this species is more widely distributed in New Guinea than in Australia. Its full diet is not known but it presumably is insectivorous, although it is known to take nectar from flowers. It sleeps during the day in a spherical nest of leaves which may be in a tree-hole or a variety of other situations. Several individuals may share a nest. Although tropical, it may become torpid during the day.

The breeding season extends from August to February, with peaks in early spring and summer. The female has four teats and rears up to four young, which are left in the nest after they vacate the pouch.

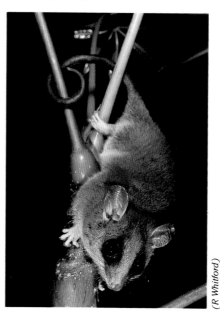

(R Whitford)

HABITAT: upland tropical rainforest
HEAD AND BODY: 10–11 cm
TAIL: 13–15 cm
DISTRIBUTION: 10,000–30,000 km^2
ABUNDANCE: common
STATUS: probably secure

Western Pygmy-possum

Cercartetus concinnus (kon-sin'-us: "elegant *Cercartetus*")

Although about the same length as the Eastern Pygmy-possum, the western species is more slightly built. It is an active climber among shrubs, using the prehensile tail as a fifth limb. It also descends to the ground, and it is frequently caught when pitfall traps are set. It feeds mainly on insects, caught in the mouth but held with the forepaws while being eaten.

It sleeps during the day, usually in a spherical nest of leaves in a small tree-hole or similar cavity. Sleep is usually associated with torpor, the body temperature falling almost to that of the surroundings.

Breeding takes place throughout the year. The female has six teats and six young may be raised. These complete their development in the nest after vacating the pouch. Two, sometimes three, litters may be raised in a year.

HABITAT: heathland to woodland and dry sclerophyll forest with dense understorey
HEAD AND BODY: 7–11 cm
TAIL: 7–10 cm
DISTRIBUTION: 300,000–1 million km^2
ABUNDANCE: common
STATUS: secure

(CL Gill)

Little Pygmy-possum

Cercartetus lepidus (lep'-id-us: "scaly [-tailed] *Cercartetus*")

(CA Henley)

The common name of this species is apt: weighing less than 10 grams, the Little Pygmy-possum is the smallest of the living diprotodont marsupials. An active predator of insects, spiders and small lizards, it inhabits the shrubby understorey, where it climbs with agility, using the tail as a fifth limb.

The nest is a simple structure of strips of bark which may be lodged in a small tree-hole, under loose bark, in an acute-angled fork, in an artefact or even on the forest floor. It sleeps in the nest during the day and, like other burramyids, frequently becomes torpid. Being the smallest of the pygmy-possums and

having the southernmost distribution, it has the greatest need to conserve energy. Indeed, it is puzzling to find the smallest pygmy-possum in Tasmania: one might have expected the opposite.

Breeding takes place from August to December. The female has four teats in a shallow pouch and usually rears four young, which are left in the nest after they vacate the pouch. Only one litter is reared in a year.

HABITAT: wet sclerophyll forest to semiarid mallee country
HEAD AND BODY: 5–7 cm
TAIL: 6–8 cm
DISTRIBUTION: 30,000–100,000 km²
ABUNDANCE: sparse
STATUS: probably secure

Eastern Pygmy-possum

Cercartetus nanus (nah'-nus: "dwarf *Cercartetus*")

(R Whitford)

Unlike other burramyids, the Eastern Pygmy-possum feeds · mainly on nectar and pollen, obtained from the flowers of a wide variety of trees and shrubs by means of a long brush-tipped tongue. Insects and soft fruits are also eaten. During the warmer part of the year, when food is plentiful, it accumulates body fat and the base of the tail becomes swollen and carrot-shaped.

It rests by day in a spherical nest of shredded bark in a small tree-hole or under the loose bark of a tree. In winter, when food is scarce, it conserves energy by becoming torpid for much of the day.

The breeding season is from August to April and two litters are usually raised in a year. The female has four or five teats and usually raises four young. After vacating the pouch, the young remain in the nest

until weaned at the age of about two months.

HABITAT: subtropical rainforest to woodland and heath
HEAD AND BODY: 7–11 cm
TAIL: 8–11 cm
DISTRIBUTION: 300,000–1 million km²
ABUNDANCE: common
STATUS: secure

Superfamily PETAUROIDEA
(pet'-or-oy'-day-ah: "*Petaurus*-superfamily" after a genus of gliders)

Throughout most of the twentieth century, the ringtail possums and gliders were placed in a single family, the Petauridae. In 1983 the ringtails (and the Greater Glider) were separated into the family Pseudocheiridae, in recognition of significant differences between the two groups. However, they are much more closely related to each other than to any other possums and, in recognition of this, they are grouped together in a superfamily.

Family PETAURIDAE
(pet-or'-id-ee: "*Petaurus*-family")

This group is restricted to the smaller furry-tailed gliders of the genus *Petaurus*, the closely related Leadbeater's Possum (*Gymnobelideus*) and, with less certainty, the striped possums. All petaurids are arboreal and most species inhabit well-watered forests. A little more than half of the species occur in New Guinea and three are restricted to that island. However, many species occur in south-eastern Australia and one extends into Tasmania.

The gliding petaurids resemble the Greater Glider but are readily distinguishable in that their gliding membranes extend from wrist (not elbow) to ankle: no difference has yet been observed in their mode of flight. The striped possums have several specialisations for feeding on wood-boring insects, and in the structure and use of their incisors and forefeet they bear a strong resemblance to the Aye-aye, a primitive primate from Madagascar. This is an example of convergent evolution, no more nor less remarkable than the similarity between gliding petaurids and the so-called flying squirrels, which are rodents.

Female petaurids have a well-developed, forward-opening pouch, enclosing two teats.

In contrast to pseudocheirids, which are mostly leaf-eaters, petaurids feed on insects and/or plant exudates (gum and sap), nectar and pollen.

Genus Dactylopsila
(dak'-til-op'-sil-ah: "naked-finger")

The name of this genus refers to the sparse fur on the digits of striped possums. Four species of striped possums occur in New Guinea and one of these also occurs in northern Queensland rainforests. The genus is distinguished by its vivid stripes of black and white, a very elongate fourth finger, and very powerful lower incisors.

Striped Possum
Dactylopsila trivirgata (trie'-ver-gah'-tah: "three-striped naked-finger")

(L J Roberts)

long thin fourth finger. During the day, it sleeps in a leaf-lined tree-hole.

Breeding takes place from February to August; mating involves a vigorous and very noisy struggle. The female usually rears two young.

HABITAT: tropical rainforest, adjacent sclerophyll forest and woodland
HEAD AND BODY: 26–27 cm
TAIL: 31–34 cm
DISTRIBUTION: 30,000–100,000 km^2
ABUNDANCE: very sparse
STATUS: probably secure

In Australia, this species is referred to simply as the "Striped Possum": to distinguish it from the other three species in the genus, it should be called the "Common Striped Possum". It is readily distinguished from other Australian mammals by its skunk-like pattern of longitudinal black and white stripes, its offensive odour and its peculiarly elongated fourth fingers. The first two of these advertisements probably serve as a warning that its flesh is unpalatable; the third is a "winkle-picking" device.

The Striped Possum is seldom seen but often heard. It runs along limbs and often hurls itself off the end of one to fall into foliage below with a resounding crash. It feeds largely on the larvae of wood-boring insects which it uncovers by gouging into a limb or branch with its long, sharp lower incisors; the insect is removed with its long tongue or the

Genus Gymnobelideus

(jim'-noh-bel-id'-ay-us: "naked-*Belideus*")

The name of this genus is intended to imply that it is like a glider (*Belideus* is an old name for *Petaurus*) but lacks a gliding membrane. This is a reasonable short definition of the genus.

Leadbeater's Possum

Gymnobelideus leadbeateri (led'-beet-er-ee: "Leadbeater's naked-*Belideus*"; J.Leadbeater was assistant to F. McCoy, describer of the genus and species)

In almost every respect except its slightly smaller size and the absence of scent glands and a gliding membrane, this species closely resembles the Sugar Glider. It is a rare species, restricted to a small area of Victoria. Mountain Ash is a tall eucalypt which drops its lower branches and forms a rather dense high canopy, and it is in this high foliage that Leadbeater's Possum lives. It seems to have evolved from a gliding petaurid very similar to the Sugar Glider but to have lost the gliding membrane because the gliding habit was of no value—perhaps even a disadvantage—in a high, continuous canopy.

Like the Sugar Glider, it feeds on eucalypt and acacia gums and on insects, but it does not make incisions in trunks to encourage the flow of sap. It nests communally in leaf-lined tree-holes but is monogamous rather than poly-gamous.

Breeding occurs through most of the year but with peaks in May and June and from October to December. The female usually rears two young, which remain in the nest after vacating the pouch.

(PR Brown)

HABITAT: tall montane Mountain Ash forest
HEAD AND BODY: 15–17 cm
TAIL: 15–18 cm
DISTRIBUTION: less than 10,000 km^2
ABUNDANCE: rare
STATUS: vulnerable

Genus Petaurus

(pet-or'-us: "rope-dancer")

The name of this genus reflects its agility. *Petaurista* is a Greek word meaning a rope-dancer, tumbler or acro-bat. Members of the genus are fluffy-tailed and have a gliding membrane extending from the wrist (or fifth digit) to the ankle.

Yellow-bellied Glider

Petaurus australis (os-trah'-lis: "southern rope-dancer")

Largest of the petaurids, this species weighs about half a kilogram, its thick fur making it appear larger than such a weight would suggest. It feeds predominantly on nectar and pollen from eucalypt blossoms and sap that exudes from incisions gnawed into the trunks of eucalypt trees. These components of the diet satisfy its need for carbohydrates, but most of the protein in the diet comes from insects obtained by lifting the bark of trees.

During the day, the Yellow-bellied Glider sleeps in a leaf-lined nest in a hole in a tall tree. At night, it climbs through the foliage and glides from tree to tree in the same manner as the Greater Glider. It is a social species, moving about in groups consisting of a male, several females and their young.

Breeding extends through most of the year except late autumn and early winter. The female has a pouch which is partially divided into left and right compartments. Only one young is reared; after vacating the pouch, it remains in the nest for about two months.

HABITAT: tall eucalypt forest
HEAD AND BODY: 27–30 cm
TAIL: 42–48 cm
DISTRIBUTION: 300,000–1 million km²
ABUNDANCE: sparse
STATUS: probably secure

(R Whitford)

Sugar Glider

Petaurus breviceps (bre'-vee-seps: "short-headed rope-dancer")

This small glider (weighing less than 150 grams) has an extensive distribution over most of the reasonably well-watered parts of Australia except the south-western corner. It feeds on the gum exuded by acacias and eucalypts and the sap exuded from long incisions made in the bark of eucalypts. Its protein requirements are provided by a wide range of arboreal insects and in winter, when these are hard to find, Sugar Gliders lose condition. The gliding membrane extends from the fifth finger to the ankle but the mode of flight appears to be the same as that of the Greater Glider.

During the day the Sugar Glider sleeps in a leaf-lined nest in a tree-hole. A number of adults and their dependent young may share a nest, possibly as a means of conserving body heat. Individuals may enter a temporary state of torpor when food is scarce, thus economising on energy resources.

Most breeding takes place from July to November. The female usually rears two young. These are left in the nest after they vacate the pouch.

(G Suckling)

HABITAT: wet and dry sclerophyll forest and woodland, usually with acacia understorey
HEAD AND BODY: 16–21 cm
TAIL: 17–21 cm
DISTRIBUTION: more than 1 million km²
ABUNDANCE: common
STATUS: secure

Squirrel Glider

Petaurus norfolcensis (nor'-foh-ken'-sis: "Norfolk Island rope-dancer")

(CA Henley)

The scientific name of this species perpetuates an error in labelling, which indicated that the first specimen to be described came from Norfolk Island. The Squirrel Glider does not occur outside the Australian mainland. It is barely distinguishable from the Sugar Glider except in being almost twice as heavy and having a somewhat more pointed snout and more defined markings. The behaviour of the two species is almost identical and they have interbred in captivity, producing fertile offspring. There is thus a strong possibility that one is a variety of the other, in which case, the name *Petaurus breviceps*, dating from 1838, would be subsumed by *Petaurus norfolcensis*, dating from 1792 (when the animal was considered to be a flying squirrel).

The major difference between the Squirrel Glider and the Sugar Glider is that the former occupies a drier habitat on the continental side of the Dividing Range.

HABITAT: dry sclerophyll forest and woodland
HEAD AND BODY: 18–23 cm
TAIL: 22–30 cm
DISTRIBUTION: 300,000–1 million km²
ABUNDANCE: common
STATUS: probably secure

Family PSEUDOCHEIRIDAE

(sue'-doh-kie'-rid-ee: "*Pseudocheirus*-family")

The Pseudocheiridae—ringtails and the Greater Glider—are represented in the well-watered forests of Australia and also in New Guinea and adjacent islands.

As their name indicates, typical ringtails (*Pseudocheirus*) have a strongly prehensile tail, which is bare for some distance on the underside of the tip; hair on the tail is much shorter than in brushtails. The Rock Ringtail, which is active on the ground, has a shorter and less prehensile tail. The Lemuroid Ringtail (*Hemibelideus*) has a bushy tail which is moderately prehensile. The Greater Glider (*Petauroides*) has a furry tail which is much longer than the body and is slightly prehensile.

There are three pairs of upper incisors but only one lower pair; upper canines are present but there are no lower ones. Leaves form the basic diet of all pseudocheirids. The grinding teeth are furnished with a number of crescentic cusps which provide an excellent mechanism for grinding tough leaves into small particles. The Greater Glider feeds exclusively on eucalypt leaves and, apart from the Koala, is the only mammal able to do so.

All pseudocheirids are excellent climbers but, as already mentioned, the Rock Ringtail spends much time on the ground. All species have the ability (shared with the Koala, cuscuses and the Scaly-tailed Possum) to oppose the first two fingers on the forefoot against the other two, and all have a well-developed first toe on the hind foot, opposable to the other toes.

All female pseudocheirids have a well-developed, forward-opening pouch which encloses two or four teats.

Recent research indicates that *Pseudocheirus* is in an active stage of evolution, with several populations on the way to becoming a new species. Relationships between *Pseudocheirus* and the other pseudocheirids are not clear. The Lemuroid Ringtail (*Hemibelideus*) appears to be rather closely related to the Greater Glider (*Petauroides*).

Genus Hemibelideus

(hem'-ee-bel-id'-ay-us: "half-*Belideus*"; *Belideus* is a name once given to fluffy-tailed gliders of the genera *Petaurus* and *Petauroides*)

The name of this genus refers to some anatomical similarities between it and the Greater Glider. It is distinguished by its rather flat face and large eyes.

Lemuroid Ringtail Possum

Hemibelideus lemuroides (lee'-mer-oy-dayz: "lemur-like half-*Belideus*")

Called "lemur-like" because of its flat face and large, forward-directed eyes, this species is only slightly larger than the Common Ringtail. Its tail is well furred and only the extreme tip

(M Trenerry)

is bare. It seems that the tail is used less as a fifth limb than as a balancer during leaping: it is a characteristic of the Lemuroid Ringtail that it launches itself into the air from the tip of a branch and, with limbs spread and tail extended, lands in the foliage of a lower limb. This species is closely related to the Greater Glider and it is tempting to see its behaviour as a forerunner to gliding.

It is restricted to an area of less than 3000 square kilometres of upland rainforest in northern Queensland where it feeds mainly on leaves from a wide variety of trees but also eats some flowers and fruits. It sleeps by day in a nest in a tree-hole.

Breeding takes place from about July to October. The female has two teats but usually rears only one young, which is carried on the mother's back after leaving the pouch.

HABITAT: upland tropical rainforest
HEAD AND BODY: 31–35 cm
TAIL: 33–37 cm
DISTRIBUTION: 10,000–30,000 km^2
ABUNDANCE: very sparse
STATUS: vulnerable

Genus Petauroides

(pet'-or-oy'-dayz: "*Petaurus*-like")

The name of this genus draws attention to a superficial similarity between the Greater Glider, *Petauroides volans*, and the other fluffy-tailed gliders of the genus *Petaurus* (members of the family Petauridae). One notable difference is that the gliding membrane of *Petauroides* extends only to the elbow, whereas it extends to the wrist in *Petaurus*.

Greater Glider

Petauroides volans (voh'-lanz: "flying *Petaurus*-like")

Weighing as much as 1.7 kilograms, this species deserves its common name (the Yellow-bellied Glider, next largest of the volplaning marsupials, weighs less than half a kilogram). Launching itself from a high tree, the Greater Glider extends a flap of skin between the flanks, elbows and ankles and glides on this squarish membrane to the trunk of another tree, steering with the long tail and by alteration of the curvature of the membrane on either side of its body. As it approaches the target tree, it swoops upward and stalls, with feet and claws outstretched, to grasp the trunk.

It feeds exclusively on eucalypt leaves. During the day it sleeps in a hole in an old, tall tree, and its distribution appears to be crucially limited by the presence of such nesting sites. Coloration is very variable, ranging from dark brown to creamy white above and white below.

The Greater Glider is solitary but pairs may share a nest from about February to October, mating in March or May. The female has two teats in a forward-directed pouch but usually raises only one young. After quitting the pouch, a young glider remains in the nest for up to four months. It may be carried on the mother's back while she is climbing but not while she is in flight.

HABITAT: wet and dry sclerophyll forest to woodland
HEAD AND BODY: 35–45 cm
TAIL: 45–60 cm
DISTRIBUTION: 300,000–1 million km²
ABUNDANCE: sparse
STATUS: probably secure

(*E Beaton*)

Genus *Pseudocheirus*

(sue'-doh-kie'-rus: "false-hand")

This genus includes the ringtail possums, which are medium-sized phalangeroids with a strongly prehensile tail and the ability to oppose the first two digits of the hand against the other three. The major component of their diet is leaves.

The Common Ringtail Possum is known to produce two types of faeces. One type, partially digested in the large caecum, is reingested and passes through the gut a second time, emerging as the "final" faeces. This device, which may well be employed by other members of the genus, improves the digestion of the fibrous leaves which make up the bulk of the diet.

Green Ringtail Possum

Pseudocheirus archeri (ar'-cher-ee: "Archer's false-hand"; the Archer family, living near Rockhampton, befriended C. Lumholtz, describer of the species)

Like the Herbert River and Lemuroid Ringtails, this species is restricted to northern Queensland. It lives mainly in the canopy and subsists entirely on leaves of a wide variety of rainforest trees. As the common name implies, the fur has an unusual green tinge. (Some authorities place this species in a separate genus, *Pseudochirops.*)

It is a large ringtail with a plump body and a rather thick but powerful prehensile tail, used as a fifth limb as the animal climbs swiftly through the canopy. Like the other ringtails, it is nocturnal, but it seems not to build a nest: during the day, it sits on a branch, curled up into a ball.

Most breeding probably occurs in June or July. The female has two teats but usually rears only one young, which is carried on the mother's back for some time after it leaves the pouch.

HABITAT: upland tropical rainforest
HEAD AND BODY: 34–38 cm
TAIL: 31–33 cm
DISTRIBUTION: 10,000–30,000 km^2
ABUNDANCE: sparse
STATUS: vulnerable

(*Y Dymock*)

Rock Ringtail Possum

Pseudocheirus dahli (dah'-lee: "Dahl's false-hand", after K. Dahl, who collected the first specimens)

Like the Scaly-tailed Possum (to which it is not closely related), the Rock Ringtail sleeps by day in the shelter of rocks and emerges at night to feed in trees, a strategy appropriate to a habitat where few trees grow large enough to provide holes sufficiently large to accommodate a possum weighing as much as 2 kilograms. The distribution of the Rock Ringtail is therefore limited to areas with rock-piles, large boulders or deep fissures and with open woodland, open forest or vine-forest.

In comparison with the arboreal ringtails, it has a longer snout and shorter tail, legs and claws. It feeds on leaves, flowers and fruits of a wide variety of trees and shrubs. Although apparently a solitary species, it may form aggregations where food is plentiful.

Females are larger than males. Breeding occurs throughout the year. The female has two teats but normally rears only one young, which is carried on the mother's back for a period after leaving the pouch and subsequently follows her on foot until it is independent.

HABITAT: rocky outcrops in tropical woodland
HEAD AND BODY: 33–38 cm
TAIL: 20–27 cm
DISTRIBUTION: 100,000–300,000 km^2
ABUNDANCE: sparse
STATUS: probably secure

(I. Morris)

Herbert River Ringtail Possum

Pseudocheirus herbertensis (her'-bert-en'-sis: "Herbert [River] false-hand")

(GA & MM Hoye)

construct dreys like those of the Common Ringtail, in response to a lack of tree-holes.

The female has two teats and commonly rears two young. After leaving the pouch they may be carried for some days on the mother's back but thereafter they are left in the nest until they become independent.

HABITAT: upland tropical rainforest
HEAD AND BODY: 30–38 cm
TAIL: 33–40 cm
DISTRIBUTION: 10,000–30,000 km^2
ABUNDANCE: very sparse
STATUS: vulnerable

Restricted to northern Queensland, this species occurs in two geographically distinct forms which differ in appearance. The northern form is light brown above and pale below; the southern form is brownish black above and white below. At present, the northern form is regarded as a subspecies of *Pseudocheirus* *herbertensis*, but it may prove to be a separate species.

The Herbert River Ringtail is a careful climber which uses its long, narrow tail as a fifth limb. It feeds mainly on leaves of rainforest trees, supplemented by fruits. During the day it usually sleeps in a nest in a tree-hole, but it has been known to

Common Ringtail Possum

Pseudocheirus peregrinus (pe'-re-green'-us: "foreign false-hand")

Smallest of the ringtails, this species has the widest distribution. It occurs in eastern and western Australia and is the only petaurid to inhabit Tasmania and some Bass Strait islands. It feeds on eucalypt leaves, buds, blossoms and soft fruits. An agile climber, it uses its long white-tipped tail as a fifth limb.

Typically, it sleeps by day in a leafy nest in a tree-hole, but where such cavities are not available it constructs a spherical drey of shredded bark. It is not very aggressive: several individuals may share overlapping home ranges and nests or dreys may be in close proximity.

Breeding takes place from April to November. The female has four teats but normally rears two young, which travel on the mother's back after they leave the pouch.

(*P German*)

HABITAT: rainforest to shrubby woodlands, suburban gardens
HEAD AND BODY: 30–35 cm
TAIL: 30–35 cm
DISTRIBUTION: more than 1 million km²
ABUNDANCE: abundant
STATUS: secure

Superfamily TARSIPEDOIDEA

(tar'-sip-ed-oy'-day-ah: "*Tarsipes*-superfamily")

Tarsipes, the Honey Possum, has long been recognised as a very peculiar "possum", highly specialised for its diet of nectar and also differing from the other possums in many aspects of its anatomy. At a time when the other possums were lumped together in one superfamily, *Tarsipes* was separated into a superfamily of its own. However, recent immunological studies provide strong evidence that the Honey Possum and the feathertails are much more closely related to each other than either is to the other "possums". In recognition of this, the feathertails (Acrobatidae) and Honey Possum (Tarsipedidae) are now placed together in the Tarsipedoidea— possibly the most radical reorganisation in the history of marsupial classification.

Family ACROBATIDAE

(ak'-roh-bah'-tid-ee: "*Acrobates*-family")

This family, which until recently was included within the Burramyidae, contains only two species, the Feathertail Glider from Australia and the larger, non-gliding Feathertail Possum from New Guinea. Acrobatids are characterised by a long thin tail bearing stiff lateral hairs that form a feather-like structure. The structure is much better developed in the Feathertail Glider than in the Feathertail Possum. Since the feather-like tail is an adaptation to gliding, it would seem that the Feathertail Possum is a non-gliding descendant of a gliding ancestor, similar in this respect to Leadbeater's Possum.

Genus Acrobates

(ak'-roh-bah'-tayz: "acrobat")

The characteristics of this genus are those of the single species. It is distinguished from all other Australian mammals by the feather-like structure of its tail.

Feathertail Glider

Acrobates pygmaeus (pig-mee'-us: "pygmy acrobat")

(CA Henley)

Anyone who has made paper aeroplanes is aware that smaller ones are more erratic in flight. On similar considerations, there is probably a lower limit to the efficient size of a gliding mammal and, with a head-and-body length of about 75 millimetres and a weight of less than 15 grams, the Feathertail Glider is probably near that limit. Its gliding membrane is also proportionately the smallest in any gliding marsupial, extending from the elbow to the knee, but it can glide for up to 300 times its head and body length, using the tail as a steering aerofoil. Quite remarkably, the pads on its toes are finely grooved, enabling it to support itself upside-down from a sheet of glass, like some geckos.

It forages with swift movements at all levels of the forest, including the understorey, feeding on small invertebrates found under the bark of trees and taking nectar, pollen and sap from eucalypts, banksias and mint-bushes. During the day it sleeps, often communally, in a spherical nest of leaves located in a tree-hole or a variety of other small spaces.

The female has four teats in a well-developed, forward-opening pouch, but normally rears two or three young, which are left in the nest after vacating the pouch. Two litters may be reared in a year.

HABITAT: cool-temperate to tropical wet and dry sclerophyll forest, extending into woodland
HEAD AND BODY: 6–8 cm
TAIL: 7–8 cm
DISTRIBUTION: more than 1 million km^2
ABUNDANCE: sparse
STATUS: secure

Family TARSIPEDIDAE

(tar'-sip-ed'-id-ee: "*Tarsipes*-family")

The characteristics of the family are those of the single species.

Genus Tarsipes

(tar'-si-pez: "Tarsier-foot")

The single species is distinguished by a very reduced dentition; a long, brush-tipped tongue; and by the lack of typical claws: these are nail-like and lie on the upper tips of the digits (except the conjoined second and third digits of the hind foot, which bear grooming claws).

Honey-possum

Tarsipes rostratus (ros'-trah'-tus: "[long-] snouted Tarsier-foot")

(RL Smith)

The Honey-possum is a small mammal which feeds on nectar and pollen by means of a long brush-tipped tongue. In the course of its evolutionary specialisation, the teeth have been reduced to a pair of pointed lower incisors and a variable number of peg-like rudiments in the upper and lower jaws, equivalent to the canines and grinding teeth of other diprotodonts. Being completely dependent upon nectar and pollen, it is restricted to areas of mixed vegetation in which some plants are in flower at any time of the year, a requirement which is met by species of the family Proteaceae in heaths of south-western Australia. The Honey-possum climbs in this dense vegetation, gripping branches and twigs with its long prehensile tail and the pads at the ends of its fingers and toes (its claws being reduced to nail-like structures like those of a tarsier). It frequently moves about on the ground and is readily caught in pit-traps. It sleeps by day, often communally, in any available shelter, including abandoned bird nests. It often becomes torpid.

Males may be longer than females, but females may be heavier. Breeding takes place throughout the year but least in midsummer. The female has four teats in a forward-directed pouch but usually rears only two or three young. At birth, these weigh 3–6 milligrams and are less than 2 millimetres long—the smallest known newborn mammals. The young are well furred when they vacate the pouch and may ride on the mother's back for a few days before being left in the nest for the remaining period of suckling. Several litters may be reared in a year.

HABITAT: cool-temperate sandplain heathland
HEAD AND BODY: 4–10 cm
TAIL: 5–11 cm
DISTRIBUTION: 100,000–300,000 km^2
ABUNDANCE: sparse
STATUS: probably secure

Superfamily MACROPODOIDEA

(mak'-roh-pod-oy'-day-ah: "*Macropus*-superfamily")

This large group (59 living species) includes all the kangaroos and their relatives. The name refers to the large hind legs with long feet upon which they can hop at considerable speed. It is also characteristic of kangaroos and most wallabies that they eat grass: with the exception of the wombats (and perhaps the Pig-footed Bandicoot), they are the only grazing marsupials. The group also includes the largest living marsupials. It is understandable that they should be popularly regarded as the "typical" marsupials, but they are in fact among the most highly specialised.

All female macropodoids have a well-developed, forward-opening pouch which can accommodate a young animal ("joey") long after it has detached itself from the teat and is capable of independent movement and feeding. Except in the Musky Rat-kangaroo (which usually has twins), only a single young is born. The pouch encloses four teats.

The superfamily Macropodoidea comprises two families. The Potoroidae include the small and rather unfamiliar rat-kangaroos, potoroos and bettongs, which retain some evidences of their phalangeroid ancestry. Within the Potoroidae we recognise two subfamilies, the Hypsiprymnodontinae, which accommodates the very interesting Musky Rat-kangaroo; and the Potoroinae, which includes all the potoroos and bettongs. The family Macropodidae includes all the kangaroos, wallabies, rock-wallabies and tree-kangaroos and some other less familiar big-footed marsupials. It is divided into two subfamilies: the Sthenurinae, which has many fossil representatives but only one living species, the Banded Hare-wallaby; and the Macropodinae, containing the remaining 48 species. According to normal practice we would refer to all members of the superfamily Macropodoidea as macropodoids but, in common usage, this term has been abbreviated to "macropods". A macropodid is a member of the family Macropodidae and a macropodine is a member of the subfamily Macropodinae. We must be careful to avoid making generalisations about macropods based on what kangaroos and wallabies do, for these macropodines represent only one of the four subfamilies of the Macropodoidea.

Family POTOROIDAE
(pot'-oh-roh'-id-ee: "*Potorous*-family")

Although at least one fossil potoroid was the size of a kangaroo, the living species range from about 40 to 80 centimetres in total length and from about 400 grams to 2 kilograms in weight (a male Eastern Grey Kangaroo can exceed 3 metres in length and weigh nearly 70 kilograms). Potoroids eat relatively non-fibrous foods such as insects and other invertebrates, bulbs, tubers and fungi; the Rufous Bettong is the only species to eat significant amounts of green plant material. They have three pairs of upper incisors, arranged in a deep arch on the upper jaw, and a pair of long lower incisors that bite against these. All have small upper canines which, like the vestigial second lower incisors often present in the Musky Rat-kangaroo, serve to remind us of the possum ancestry of the macropods. The stomach is not divided into compartments.

The hind limbs are not as disproportionately large nor the feet as long as those of macropodids. Most potoroids hop when moving fast, but may also employ the forelimbs, particularly when turning: the Musky Rat-kangaroo is unique among the macropods in that it does not hop. The tail is slightly prehensile and is used to carry nesting material: it is never employed, as in typical macropodids, to support the body while the hind legs are being swung forward in slow locomotion.

Subfamily HYPSIPRYMNODONTINAE
(hip'-see-prim'-noh-dont'-in-ee: "*Hypsiprymnodon*-subfamily")

This subfamily contains only one living species, the Musky Rat-kangaroo. Had it not been discovered, zoologists would have "invented" something like it to bridge the evolutionary gap between possums and macropods. Unlike all the other macropods, it retains a first digit on the hind foot, set at right angles to the axis of the foot as in arboreal marsupials. Otherwise, the hind limb conforms to the typical macropod pattern but the disproportion between it and the forelimb is no greater than in ringtail or brushtail possums.

Genus Hypsiprymnodon
(hip'-see-prim'-noh-don: "potoroo-tooth")

Hypsiprymnus was a name once given to the potoroos: the name therefore means "potoroo-toothed", referring to the large cutting premolar tooth.

Musky Rat-kangaroo
Hypsiprymnodon moschatus (mos-kah'-tus: "musky potoroo-tooth")

(L. Robinson)

Smallest of the macropods, the Musky Rat-kangaroo is known only from northern Queensland rainforest, where it feeds among the leaf litter on insects, other invertebrates and fallen fruits. It is active in the early morning and late afternoon but sleeps at night and through the middle of the day in a substantial nest of vegetation, often placed between the bulbous roots of a rain-forest tree. On the ground, it proceeds quadrupedally by a "bunny hop", extending the forelegs then bringing both hind legs forward. Its fast locomotion is an extension of this gait into a series of bounds: it does not hop.

At least in juvenile animals, the mobile first toe of the hind foot is used to grip branches when climbing in undergrowth. The moderately pre-hensile, scaly tail is used to carry nesting material.

Sexual maturity is reached at about 13 months and mating takes place from February to July. The female has four teats in a forward-opening pouch and usually rears two young, which are left in the nest after they vacate the pouch.

HABITAT: tropical rainforest
HEAD AND BODY: 15–27 cm
TAIL: 12–16 cm
DISTRIBUTION: 30,000–100,000 km^2
ABUNDANCE: very sparse
STATUS: probably secure

Subfamily POTOROINAE

(pot'-oh-roh-een'-ee: "*Potorous*-subfamily")

The Potoroinae comprise all the potoroids except *Hypsiprymnodon*.

Genus Aepyprymnus

(ee'-pi-prim'-nus: "high-rump")

The name of this genus indicates that, when the animal is resting or feeding, the hindquarters are higher than the rest of the body (as in all macropods). *Aepyprymnus* is closely related to *Bettongia* and is generally referred to as one of the bettongs. It can be distinguished by having hairs on the central parts of the muzzle (naked in the other potoroines).

Rufous Bettong

Aepyprymnus rufescens (rue-fes'-enz: "reddish high-rump")

The Rufous Bettong is unusual among potoroids in habitually eating grasses, sedges and herbs. It also forages for tubers in the usual manner of bettongs. During the day it sleeps in a conical nest of grass built over a shallow scrape, usually in the shade of a shrub or tussock. Nesting material is carried to the site with the prehensile tail.

Females are larger than males. Sexual maturity is reached at about 12 months. Breeding takes place throughout the year. The female rears only one young, which follows the mother for almost two months after quitting the pouch.

HABITAT: dry sclerophyll forests with dense understorey
HEAD AND BODY: 37–39 cm
TAIL: 34–39 cm
DISTRIBUTION: 300,000–1 million km²
ABUNDANCE: common
STATUS: probably secure

(LF Schick)

Genus Bettongia

(bet-ong'-gee-ah: "bettong")

Bettongs have shorter, broader heads than potoroos, their bodies are somewhat stouter and their feet are proportionately longer (hind foot longer than the head). The tail is about as long as the head and body. Some bettongs live in well-watered forested areas; others live, or lived, in more arid areas. The latter have become rare or extinct.

Tasmanian Bettong

Bettongia gaimardi (gay-mar'-dee: "Gaimard's bettong", after J. P. Gaimard, French naturalist)

(*RW Rose*)

The species that is now called the Tasmanian Bettong once had an extensive distribution in south-eastern Australia on the coastal side of the Dividing Range; it is now extinct on the mainland. It appears to be closely related to the Brush-tailed Bettong but is somewhat larger and has proportionately longer hind feet. It forages at night, digging with its long-clawed, powerful forelimbs for fungi, tubers and bulbs. By day it sleeps in a well-constructed nest of grass and bark, situated under a shrub, tussock or fallen limb.

Sexual maturity is reached at the age of one year. Breeding takes place throughout the year. The female rears only one young, which follows the mother after it leaves the pouch. The Tasmanian Bettong exhibits embryonic diapause.

HABITAT: temperate dry sclerophyll forest with grassy understorey; grassland
HEAD AND BODY: 31–33 cm
TAIL: 28–35 cm
DISTRIBUTION: 10,000–30,000km²
ABUNDANCE: common
STATUS: probably secure

Burrowing Bettong

Bettongia lesueur (le-swer': "Le Sueur's bettong", after C. A. Le Sueur, French naturalist)

Slightly larger than the Brush-tailed Bettong, this species is remarkable among the macropods for constructing burrows several metres long, at the end of which is a nest of vegetation. Burrows are often grouped together to form a warren accommodating scores of individuals. At night, the Burrowing Bettong forages in much the same manner as other bettongs, feeding on tubers, bulbs, fungi and insects, including termites. It also eats green plants, including peas and beans in kitchen gardens, and carrion. Within historical times its range has decreased from the south-western half of Australia to four small islands off the coast of Western Australia.

Sexual maturity is reached at five months. Breeding takes place throughout the year. The female rears only one young.

HABITAT: tropical to temperate sclerophyll forest to spinifex grassland
HEAD AND BODY: 28–40 cm
TAIL: 22–30 cm
DISTRIBUTION: less than 10,000 km²
ABUNDANCE: common
STATUS: vulnerable

(*K Johnson*)

Brush-tailed Bettong

Bettongia penicillata (pen'-is-il-ah'-tah: "brush [-tailed] bettong")

The Brush-tailed Bettong forages at night, digging with its powerful, strongly clawed forefeet for fungi, tubers, bulbs and insects. It sleeps by day in a nest of vegetation over a hollow scraped in the ground, usually in the shade of a shrub or tussock. Material for the nest is carried in the prehensile tail.

Sexual maturity is reached before the age of six months. Breeding takes place throughout the year. The female usually rears only one young, which follows its mother for several weeks after leaving the pouch and shares her nest during the day. The Brush-tailed Bettong exhibits embryonic diapause.

HABITAT: tropical to temperate dry sclerophyll forest and woodland with low shrubs or tussocks
HEAD AND BODY: 30–38 cm
TAIL: 29–36 cm
DISTRIBUTION: 10,000–30,000 km^2
ABUNDANCE: sparse
STATUS: possibly endangered

(CA Henley)

Genus Caloprymnus

(kal'-oh-prim'-nus: "beautiful rump")

The characteristics of this genus are those of the single (extinct) species. It is distinguished by a very broad face and very long feet.

Desert Rat-kangaroo

Caloprymnus campestris (kam-pes'-tris: "open-country beautiful-rump")

(J Gould)

The relationship of this species to the other bettongs is not clear. It had a very broad head and its tail and its hind feet were each longer than the combined head and body. Perhaps related to the length of its hair-fringed feet, it hopped at speed in a peculiar manner, with the right foot touching the ground in front of the left, which was inclined outwards at an angle of about 30 degrees to the line of travel. It hopped fast and long: one individual was chased nearly 20 kilometres by men galloping on horseback.

It fed at night but its diet is unknown. It slept by day in a well-constructed nest of grass and twigs. There has been no report of it since 1935, so it is probably extinct.

HABITAT: stony desert
HEAD AND BODY: 25–28 cm
TAIL: 30–38 cm
DISTRIBUTION: nil
ABUNDANCE: nil
STATUS: extinct

Genus *Potorous*

(pot'-oh-roh'-us: "potoroo", Aboriginal name)

Potoroos have a rather slender, tapering head and a short hind foot (shorter than the head). The strong forelimbs are sometimes employed in fast locomotion. The tail is well furred and prehensile and not more than about three-quarters of the length of the head and body. Potoroos tend to inhabit well-watered forests.

Long-footed Potoroo

Potorous longipes (lon'-ji-pez: "long-footed potoroo")

The Long-footed Potoroo is closely related to the Long-nosed Potoroo but weighs about twice as much. Its hind feet are, proportionately, a little longer. It appears to be similar in behaviour to the Long-nosed Potoroo, foraging at night by digging pits with its forefeet, and its diet is assumed to be similar. It sleeps by day in a nest of vegetation carried to the nest site in its prehensile tail.

Sexual maturity is reached at the age of one and a half to two years. Breeding probably takes place throughout the year.

HABITAT: wet sclerophyll forest with dense ground cover
HEAD AND BODY: 38–42 cm
TAIL: 31–33 cm
DISTRIBUTION: less than 10,000 km²
ABUNDANCE: rare
STATUS: endangered

(JB Cooper)

Broad-faced Potoroo

Potorous platyops (plat'-ee-ops: "flat-looking potoroo")

Smallest of the potoroos, this species was only a little larger than the Musky Rat-kangaroo. It had a fat-cheeked appearance. It was described in 1839 but has not been seen since 1875. Nothing is known of its biology.

HABITAT: semiarid woodland and grassland
HEAD AND BODY: about 24 cm
TAIL: about 18 cm
DISTRIBUTION: nil
ABUNDANCE: nil
STATUS: extinct

(J Gould)

Long-nosed Potoroo

Potorous tridactylus (trie-dak'-til-us: "three-toed potoroo"; refers to the structure of the typical macropod foot, with conjoined 2nd and 3rd toes, a large 4th toe and a smaller 5th toe)

The Long-nosed Potoroo is so called to contrast it with the now extinct Broad-faced Potoroo (which had a similarly tapered skull but broad cheeks). At night it uses its powerful forearms and long-clawed forefeet to dig in forest litter and soil for succulent tubers, fungi and insect larvae. When moving slowly, it progresses in a "bunny-hop"; when moving fast, it leaps on its hind legs but may also employ its forelegs, particularly when turning. By day it sleeps in a nest of vegetation carried to the nest site in the prehensile tail.

Males are slightly larger than females. Sexual maturity is reached in the first year. Breeding takes places throughout the year with a peak of births in early summer and midsummer. The female normally rears a single young, which follows the mother on foot after leaving the pouch.

(JE Wapstra)

HABITAT: cool rainforest, wet sclerophyll forest with dense ground cover, well-watered heathland
HEAD AND BODY: 32–40 cm
TAIL: 20–26 cm
DISTRIBUTION: 30,000–100,000 km^2
ABUNDANCE: sparse
STATUS: probably secure

Family MACROPODIDAE

(mak'-roh-poh'-did-ee: "*Macropus*-family")

With 48 living species, Macropodidae is the largest marsupial family. Most macropodids have the same general body shape, which is significantly departed from only in the tree-kangaroos. The shape is determined by the hopping mode of fast locomotion, which depends upon powerful hind limbs, long feet and a powerful fourth toe. The centre of gravity of a kangaroo lies above its hips, and the rather slender chest, small forelimbs and head are balanced by a long, thick-based and rather inflexible tail. There is little room for variation on this basic arrangement, which is beautifully adapted to fast movement on more or less level ground.

However, the arrangement leads to an absurdity: kangaroos have difficulty in moving *slowly*. They cannot, like the bandicoots, proceed by "bunny hops", but have to support the body on the forelimbs and the tail while swinging the hindlimbs forward—a clumsy and unique method of locomotion.

In rock-wallabies, which bound in a three-dimensional environment, the hind foot is broader, the fourth toe shorter and the tail is mobile and only slightly tapered. In the climbing tree-kangaroos, the hind foot is still broader and shorter, the forelimbs are larger and more powerful and the tail is cylindrical but less mobile than in the rock-wallabies. With these notable exceptions, most of the variation in macropodids relates to size, diet, dentition, physiology and reproduction.

The family Macropodidae comprises two subfamilies. The Sthenurinae, known mainly as fossil forms, is characterised by browsing habits and may be regarded as the more primitive group. The Macropodinae, containing most of the living species, largely comprises grazing species: the leaf-eating tree-kangaroos are a notable exception. Grazing kangaroos have a complex stomach, divided into compartments where micro-organisms can ferment fibrous plant material, rendering it more available as food. The success of the kangaroos and larger wallabies is due, in a large degree, to their ability to feed on tough, fibrous grasses.

Many kangaroos have a capacity for embryonic diapause, whereby a very early embryo can be retained in the uterus for quite a long time in a state of suspended development. In the life of a typical female macropodid, the first mating leads to an embryo which proceeds to normal development and is born. It attaches itself to a teat for a much longer period of development. Very close to the time of birth, the female mates again, but the embryo arising from this mating does not proceed past the stage of a ball of cells (blastocyst) until the first young quits the pouch. The second embryo now completes its development, is born, and attaches itself to another teat in the pouch. The female then mates a third time, resulting in another quiescent blastocyst.

The female is now suckling one young outside the pouch and another inside, while holding an embryo in reserve. Interestingly, the milk produced for the young which is attached to a teat is of quite different composition to that provided for its older sibling. This is the usual state of a typical female macropod for the remainder of her reproductive life. Should a pouch-young be lost or die from any cause, the cessation of suckling is the stimulus for the blastocyst to begin normal development: its birth is quickly followed by another mating. There has been much debate about the significance and value of embryonic diapause: it does mean that a lost joey can be replaced rapidly without waiting for mating, but the time saved is not very great.

It seems that, in general, European settlement has had a deleterious effect on most of the smaller macropods but has been beneficial to the larger species.

Subfamily MACROPODINAE

(mak'-roh-poh-deen'-ee: "*Macropus*-subfamily")

This group comprises all the kangaroos and wallabies except the Rufous Hare-wallaby. Additionally, it includes the more specialised rock-wallabies and tree-kangaroos. While most macropodines are grazing animals, the hare-wallabies are to some extent browsers and the tree-kangaroos feed largely upon the leaves of rainforest trees.

Genus Dendrolagus

(den'-droh-lah'-gus: "tree-hare")

This genus has five species in New Guinea and two in Australia, both limited to the rainforests of Cape York. They are anomalous kangaroos which have reversed some of the major trends in macropod evolution in order to gain access to the leaves of rainforest trees. The hind limbs are short and the hind feet are short and broad, with granulated soles and sharp, curved claws. The forelimbs are much larger and more powerful than in any other macropods and the disproportion between these and the hind limbs is hardly greater than in ringtail and brushtail possums. The tail, which is as much as 30 per cent longer than the head and body, is almost cylindrical and is not in the least prehensile.

A tree-kangaroo can climb a vertical trunk, clinging with the claws of its fore and hind feet. It can walk along a branch, alternating its hind feet, but when on the ground, it bounds bipedally like other macropodids. Tree-kangaroos are the only macropods able to walk backwards. In the trees, they do not seek any shelter, but in captivity they readily use nesting boxes.

Bennett's Tree-kangaroo

Dendrolagus bennettianus   (ben'-et-ee-ah'-nus: "Bennett's tree-hare", after G. Bennett, first curator of the Australian Museum)

(LJ Roberts)

Bennett's Tree-kangaroo is largely nocturnal. It moves through the rainforest canopy, feeding on leaves of trees, and sometimes descends below the canopy to feed on the leaves of vines. During the day it sits on a branch high in the canopy, with its head tucked between its knees.

Males are larger than females. Breeding is probably continuous. One young is born.

HABITAT: upland and lowland tropical rainforest
HEAD AND BODY: 50–65 cm
TAIL: 63–94 cm
DISTRIBUTION: 10,000–30,000 km²
ABUNDANCE: sparse
STATUS: vulnerable

Lumholtz's Tree-kangaroo

Dendrolagus lumholtzi (lum'-holt-zee: "Lumholtz's tree-hare", after C. Lumholtz, Norwegian naturalist who collected first specimens)

At night, this tree-kangaroo climbs through the rainforest canopy, feeding mainly on leaves but also eating fruits. During the day it sits on a branch with its head between its knees.

Males are much larger than females. Breeding is probably continuous. One young is born.

HABITAT: upland tropical rainforest
HEAD AND BODY: 48–60 cm
TAIL: 60–70 cm
DISTRIBUTION: 10,000–30,000 km²
ABUNDANCE: very sparse
STATUS: vulnerable

(R & D Keller)

Genus *Lagorchestes*

(lag'-or-kes'-tayz: "dancing-hare")

Species of *Lagorchestes* can be distinguished from *Lagostrophus* by their hairy muzzle and relatively shorter tail.

Spectacled Hare-wallaby

Lagorchestes conspicillatus (kon-spis'-il-ah'-tus: "spectacled dancing-hare")

HABITAT: tropical sclerophyll forest, woodland, shrubland, tussock grassland
HEAD AND BODY: 40–47 cm
TAIL: 37–49 cm
DISTRIBUTION: more than 1 million km²
ABUNDANCE: sparse
STATUS: probably secure

(E Beaton)

The Spectacled Hare-wallaby, so called because of the reddish rings around its eyes, feeds at night on the leaves of shrubs and spinifex. It does not need to drink. By day it sleeps in a shelter dug into the base of a large spinifex hummock. It is solitary.

Sexual maturity is reached at about 12 months. Breeding takes place throughout the year. One young is reared. The species exhibits embryonic diapause.

Rufous Hare-wallaby

Lagorchestes hirsutus (her-sue'-tus: "hairy dancing-hare")

(*K Johnson*)

are slightly larger than males.

This species once extended over most of arid and semiarid Australia but is now restricted to parts of the Tanami Desert and to Bernier and Dorre Islands in Shark Bay.

HABITAT: semiarid woodland, scrubland and grassland, particularly where regenerating after local fires
HEAD AND BODY: 31–39 cm
TAIL: 25–30 cm
DISTRIBUTION: 10,000–30,000 km²
ABUNDANCE: sparse
STATUS: vulnerable

The scientific name refers to the rather shaggy hair; the common name to its reddish colour. At night the Rufous Hare-wallaby feeds on grasses, sedges and the leaves of shrubs. In the cooler part of the year it shelters by day (like the Spectacled Hare-wallaby) in a space dug into the base of a spinifex hummock or low shrub; in hotter areas it digs a burrow less than 1 metre long under similar shelter. It probably does not need to drink. Its reproductive biology is not yet known. Females

Eastern Hare-wallaby

Lagorchestes leporides (lep'-or-ee'-dayz: "hare-like dancing-hare")

We have no knowledge of the diet or reproduction of this species, which probably became extinct towards the end of the nineteenth century, but it probably resembled the other members of the genus. During the day it slept in a shelter excavated under a large tussock, and, when disturbed from this, ran in a zigzag manner like a hare, often making prodigious leaps (as high as 1.8 metres).

HABITAT: tussock grassland
HEAD AND BODY: about 45 cm
TAIL: about 32 cm
DISTRIBUTION: nil
ABUNDANCE: nil
STATUS: extinct

(*J Gould*)

Genus *Macropus*

(mak'-roh-poos: "long-foot")

With 14 species, this genus is currently the most widespread and successful of the macropodines and, indeed, of all the macropods. Its members are all grazing species with large grinding molars and a stomach divided into compartments in which grass fibre (cellulose) is digested by micro-organisms. The forelimbs are weakly developed in females but may be very powerful in the males, which, in most species, are much larger than the females.

Agile Wallaby

Macropus agilis (a-jil'-is: "agile long-foot")

In the late afternoon and night the Agile Wallaby emerges from dense forest vegetation to graze on native grasses in open areas. During the day it sleeps, communally, in the shelter of dense vegetation. It is sociable, moving and feeding in groups of about 10, sometimes forming larger feeding aggregations.

Males are about twice the weight of females. Sexual maturity is reached at about 15 months. Breeding is continuous. One young is born. The species exhibits embryonic diapause.

HABITAT: tropical to subtropical sclerophyll forest and woodland with adjacent grassy areas
HEAD AND BODY: 59–85 cm
TAIL: 59–84 cm
DISTRIBUTION: 300,000–1 million km^2
ABUNDANCE: abundant
STATUS: secure

(F Kristo)

Antilopine Wallaroo

Macropus antilopinus (an'-til-oh-pee'-nus: "antelope-like long-foot", referring to a supposed similarity of the fur to that of an antelope)

This species is less dependent upon rocky slopes than the other wallaroos and is often found in more or less level country. It also differs from the other wallaroos in being less stockily built (more kangaroo-like) and in being gregarious, moving about in groups of three to eight individuals. It needs regular access to drinking water. During the day it sleeps in the shelter of a bush, tree or rock.

Males can be more than twice the weight of females. Breeding proceeds throughout the year, but most births are between February and June. There appears to be no embryonic diapause.

HABITAT: tropical woodland
HEAD AND BODY: 78–120 cm
TAIL: 68–90 cm
DISTRIBUTION: 300,000–1 million km^2
ABUNDANCE: common
STATUS: secure

(JE Wapstra)

Black Wallaroo

Macropus bernardus (ber-nar'-dus: "Bernard's long-foot", after Bernard Woodward, first curator of the Western Australian Museum)

As the common name implies, this species is sooty brown to black, having the darkest fur of any macropod. It grazes at night on native grasses, often descending to the plains at the foot of an escarpment. Unlike the Common Wallaroo, it needs to drink. At night it sleeps in a cave, under a rock shelf or under a low tree. It is solitary.

Males are about one and a half times the weight of females. One young is born. Nothing else is known of its reproduction.

HABITAT: steep rocky slopes with tropical woodland and grassy understorey
HEAD AND BODY: 60–75 cm
TAIL: 55–65 cm
DISTRIBUTION: 30,000–100,000 km²
ABUNDANCE: sparse
STATUS: probably secure

(I Morris)

Black-striped Wallaby

Macropus dorsalis (dor-sah'-lis: "[notably] backed long-foot", referring to prominent stripe along middle of back)

(R & A Williams)

HABITAT: tropical to warm-temperate sclerophyll forest and woodland with dense ground cover and adjacent grassy areas
HEAD AND BODY: 110–159 cm
TAIL: 54–83 cm
DISTRIBUTION: 300,000–1 million km²
ABUNDANCE: common
STATUS: secure

The Black-striped Wallaby makes runways in dense vegetation. At night it moves through these to open areas where it feeds on native grasses. During the day it sleeps communally, in the shelter of dense ground vegetation. It is gregarious, moving and feeding in groups of about 20 animals of all ages.

Males are up to three times the weight of females. Females become sexually mature at about 14 months, males at about 20 months. Breeding is continuous. A single young is born. The species exhibits embryonic diapause.

Tammar Wallaby

Macropus eugenii (yue-jay'-nee-ee: "Eugène [Island] long-foot", from l'Isle Eugène, now known as St Peter's Island, Nuyts Archipelago)

(*LF Schick*)

The Tammar Wallaby makes runways in dense vegetation but leaves these at night to feed on native grasses. It can survive for long periods without drinking, but some coastal populations drink salt water—which implies extraordinarily efficient kidneys. During the day it sleeps in the shelter of dense vegetation. It is solitary.

Males are noticeably larger than females. Females become sexually mature at nine months, males at about 22 months. A single young is born. The species exhibits embryonic diapause.

HABITAT: cool-temperate dry sclerophyll to semiarid woodland and shrubland with grassy understorey or adjacent grassy areas
HEAD AND BODY: 52–68 cm
TAIL: 33–45 cm
DISTRIBUTION: 30,000–100,000 km²
ABUNDANCE: common
STATUS: probably secure

Western Grey Kangaroo

Macropus fuliginosus (fool'-i-jin-oh'-sus: "sooty long-foot")

The Western Grey Kangaroo occupies most of the southern part of the continent, into which the Red Kangaroo barely intrudes. Like the Red Kangaroo, it is a grazer, but it requires a somewhat higher rainfall. During the day it sleeps in the shade of a tree or shrub, often communally.

It is gregarious, moving in small groups and forming much larger aggregations where food is locally abundant.

Males are about twice the size of females. Females reach sexual maturity at about 18 months, males at about two years. Breeding proceeds throughout the year. A single young is born. The species exhibits embryonic diapause.

HABITAT: dry sclerophyll forest and woodland with grassy understorey or adjacent grassy areas
HEAD AND BODY: 50–125 cm
TAIL: 42–100 cm
DISTRIBUTION: more than 1 million km²
ABUNDANCE: abundant
STATUS: secure

(*A Eames*)

Eastern Grey Kangaroo

Macropus giganteus (jie'-gan-tay'-us: "giant long-foot")

The Eastern Grey Kangaroo *is* very large, but the naturalist who bestowed its specific name thought that it was a gigantic jerboa! The biology of the Eastern Grey Kangaroo is very similar to that of the closely related western species. It feeds at night on grasses and green herbage. During the day it sleeps in the shade of a tree or shrub. It is gregarious, moving and feeding in groups of three to five individuals but sometimes forming large feeding aggregations.

Males are up to twice the size of females. Females reach sexual maturity at 18 months, males at about two years. Breeding is continuous throughout the year. A single young is born. The species exhibits embryonic diapause.

HABITAT: tropical to cool-temperate dry sclerophyll forest, woodland and shrubland, always with grassy areas
HEAD AND BODY: 50–120 cm
TAIL: 45–110 cm
DISTRIBUTION: more than 1 million km²
ABUNDANCE: abundant
STATUS: secure

(*GA Drake*)

Toolache Wallaby

Macropus greyi (gray'-ee: "Grey's long-foot", after G. Grey, explorer and governor of South Australia, who collected first specimens)

The common name of this species is pronounced toh-lay'-chee. It was abundant at the time of European settlement but became extinct in the nineteenth century, apparently because of removal of its habitat, aggravated by hunting. It fed at night on native grasses. During the day it slept in the cover of dense vegetation, often casuarina thickets. It was gregarious, moving, feeding and sleeping in groups. Nothing is known of its reproduction.

HABITAT: heathland with adjacent grassland
HEAD AND BODY: 81–84 cm
TAIL: 71–73
DISTRIBUTION: nil
ABUNDANCE: nil
STATUS: extinct

(*J Gould*)

Western Brush Wallaby

Macropus irma (er'-mah: "Irma long-foot", significance unknown, perhaps the name of a friend of M. Jourdan, the French zoologist who described the species)

The Western Brush Wallaby is associated with forests ("brushes"), but feeds in open grassy areas in the early morning or late afternoon. For the rest of the time it sleeps in the shelter of a clump of bushes or low trees. It is gregarious, moving in groups of individuals of all ages. Males and females are similar in size.

Little is known of its reproduction.

A single young is born. Most births are in April and May.

HABITAT: cool- to warm-temperate dry sclerophyll (Jarrah) forest and woodland with adjacent grassy areas
HEAD AND BODY: 90–160 cm
TAIL: 54–97 cm
DISTRIBUTION: 100,000–300,000 km²
ABUNDANCE: abundant
STATUS: probably secure

(R & A Williams)

Parma Wallaby

Macropus parma (par'-mah: "Parma long-foot", *pama* being an Aboriginal name for the species)

The Parma Wallaby makes runways in dense ground vegetation. At night it emerges to graze in open areas. During the day it sleeps in the shelter of dense vegetation. It is solitary but may form small feeding aggregations.

Males are larger than females.

Females are sexually mature at one year of age, males at about two years. Breeding is continuous, with a peak of birth from February to June. A single young is born. The species exhibits embryonic diapause.

The Parma Wallaby was introduced to New Zealand in the nineteenth century and a population persists on Kawau Island, where it is regarded as a pest of pine plantations.

HABITAT: wet montane sclerophyll forest with dense understorey and adjacent grassy areas
HEAD AND BODY: 45–53 cm
TAIL: 40–55 cm
DISTRIBUTION: 30,000–100,000 km²
ABUNDANCE: sparse
STATUS: probably secure

(R & A Williams)

Whiptail Wallaby

Macropus parryi (pa'-ree-ee: "Parry's long-foot", after E. Parry, explorer, who brought a live animal to England)

(GB Baker)

This wallaby has a rather slender tail which is a little longer than the head and body. From before dawn into the early morning and from late afternoon into the early night, it feeds on native grasses, also eating some herbs and ferns. For the remaining part of the day and night it sleeps in the shelter of a shrub or low tree. It is gregarious, moving in groups of up to 50 individuals of all ages.

Males are up to twice the weight of females. Females become sexually mature in the second year of life; males seldom mate until two to three years old. Breeding is continuous and two young may be born in a year. The species exhibits embryonic diapause.

HABITAT: wet and dry sclerophyll forest with grassy understorey or adjacent grassy areas; usually on hillsides
HEAD AND BODY: 70–95 cm
TAIL: 73–105 cm
DISTRIBUTION: 300,000–1 million km²
ABUNDANCE: abundant
STATUS: secure

Common Wallaroo (or Euro)

Macropus robustus (roh-bus'-tus: "robust long-foot")

The use of two common names for this species is justified. In eastern Australia it is a grey animal which inhabits forests and is called the Common Wallaroo; in central and western Australia it is reddish, lives in hot, arid regions and is known as the Euro. Even that statement is a simplification, for the species is divided into four subspecies, each slightly different in appearance. Like other wallaroos, it has a bare muzzle.

During the day it sleeps in shelter, usually under a rock overhang or in a cave on the upper parts of a slope. At night it descends to graze on more level ground. (Because of this habit, the damage it sometimes causes to crops or pastures is often blamed on plains-dwelling kangaroos.) It does not need to drink, obtaining all necessary water from its food. It is solitary.

Males are as much as twice the weight of females. Sexual maturity is reached between 18 and 24 months. Breeding is continuous. A single young is born. The species exhibits embryonic diapause.

HABITAT: very varied, from wet sclerophyll forest to arid tussock grassland and from tropical to sub-alpine regions; usually associated with rocky slopes with caves or rocky shelves.
HEAD AND BODY: 55–110 cm
TAIL: 53–90 cm
DISTRIBUTION: more than 1 million km²
ABUNDANCE: abundant
STATUS: secure

(P Klapste)

Red-necked Wallaby

Macropus rufogriseus (rue'-foh-griz-ay'-us: "red-grey long-foot")

(GA Hoye)

The Red-necked Wallaby grazes from late afternoon to dawn in grassy areas. It is solitary but may form feeding aggregations. It sleeps for most of the day under cover of dense vegetation.

Males are much larger than females. Females become sexually mature early in their second year; males later in the second year. Breeding is continuous on the mainland but from January to July in Tasmania. A single young is born.

The Red-necked Wallaby was introduced to New Zealand in the nineteenth century and has since become a pest.

HABITAT: subtropical to cool-temperate wet and dry sclerophyll forests to woodland, all with dense understorey and with adjacent grassy areas
HEAD AND BODY: 66–89 cm
TAIL: 62–88 cm
DISTRIBUTION: 300,000–1 million km^2
ABUNDANCE: common
STATUS: secure

Red Kangaroo

Macropus rufus (rue'-fus: "red long-foot")

(N Chaffer)

The Red Kangaroo is the dominant macropod of the drier regions of the continent. In the eastern part of its range males are usually red and females a bluish grey; elsewhere, both sexes may be reddish. Like wallaroos, the Red Kangaroo has a naked muzzle. It grazes during the night on a wide variety of grasses and low herbaceous plants; when food is scarce, it may extend its feeding into early morning and late afternoon. When water is available it will drink, but if it obtains sufficient green food it does not need to do so. It has no absolute requirement for shelter, but will lie in the shade of a bush or tree on very hot days. It is gregarious, moving in groups ranging from a few dozen to several hundred individuals.

Old males may be three times the weight of mature females. Females are sexually mature at about 18 months, males at about two years. There is a single young. The species exhibits embryonic diapause.

HABITAT: woodland, shrubland, grassland and desert from tropical to cool-temperate regions
HEAD AND BODY: 74–140 cm
TAIL: 64–100 cm
DISTRIBUTION: more than 1 million km^2
ABUNDANCE: abundant
STATUS: secure

Genus *Onychogalea*

(on'-ik-oh-gah-lay'-ah: "nailed-weasel")

Members of this genus are characterised by a flattened horny structure, rather like a thick fingernail, on the tip of the tail, the function of which remains unknown. The upper incisors are slender and decrease evenly in width from the first to the third. The claws of the forefoot are strongly developed. When a Nailtail Wallaby is hopping at speed, it holds its arms outwards and downwards from the body. The fur is fine and rather silky.

Bridled Nailtail Wallaby

Onychogalea fraenata (free-nah'-tah: "bridled nailed-weasel")

The scientific and common names of this beautiful animal refer to the white, bridle-like stripes on the head and shoulders. It feeds at night on grasses and possibly also eats succulent roots uncovered by the strongly clawed forefeet. It probably does not need to drink. During the day it retires to denser vegetation, where it sleeps in a saucer-shaped scrape in the shelter of a low tree or shrub. Males are larger than females; both sexes are solitary. The reproductive biology is unknown.

HABITAT: semiarid woodland and shrubland, both with adjacent grassy areas, often on stony ground
HEAD AND BODY: 43–70 cm
TAIL: 36–54 cm
DISTRIBUTION: less than 10,000 km²
ABUNDANCE: sparse
STATUS: endangered

(CA Henley)

Crescent Nailtail Wallaby

Onychogalea lunata (lue-nah'-tah: "crescent [moon] nailed-weasel")

The scientific and common names refer to a crescentic white stripe behind each shoulder. Although reported to have been seen in the 1950s, this species appears to be extinct. It fed at night and slept by day in a scrape under a low bush. When chased, it has been seen to take refuge in a hollow log or tree. Nothing is known of its reproductive biology.

HABITAT: semiarid to arid woodland and tussock grassland
HEAD AND BODY: 37–51 cm
TAIL: 15–33 cm
DISTRIBUTION: nil
ABUNDANCE: nil
STATUS: probably extinct

(J Gould)

Northern Nailtail Wallaby

Onychogalea unguifera (ung-wif'-er-ah: "nail-bearing nailed-weasel")

This is the largest of the nailtail wallabies (up to 9 kilograms) and the only species that is still common. It feeds at night on grasses, possibly including the roots of some species. By day it sleeps in a scrape in the soil under a low shrub. Males are much larger than females. Nothing is known of the breeding biology.

HABITAT: tropical woodland with grassy understorey; grassland
HEAD AND BODY: 50–70 cm
TAIL: 60–75 cm
DISTRIBUTION: more than 1 million km²
ABUNDANCE: common
STATUS: secure

(PM Johnson)

Genus *Peradorcas*

(pe'-rah-dor'-kas: "pouched-gazelle")

The characteristics of the genus *Peradorcas* are those of the single species. This small rock-wallaby is distinguished by having a continuous succession of molar teeth.

Nabarlek

Peradorcas concinna (kon-sin'-ah: "elegant pouched-gazelle")

(GD Sanson)

maturity is reached in the second year. Breeding takes place throughout the year, with a peak of births in the summer wet season. The species exhibits embryonic diapause.

HABITAT: rocky margins of tropical grasslands
HEAD AND BODY: 29–35 cm
TAIL: 22–31 cm
DISTRIBUTION: 100,000–300,000 km²
ABUNDANCE: very sparse
STATUS: probably secure

The Nabarlek is a very small rock-wallaby (not more than 1.5 kilograms), but it is possibly related more closely to pademelons (*Thylogale*) than to the typical rock-wallabies (*Petrogale*). It is unique among the macropods in producing a continuous succession of adult molar teeth at the rear of the jaw, replacing those that become worn down and drop out at the front of the molar row. At night (in the day during the wet season), it feeds on grasses and ferns.

During the day it sleeps in a deep crevice between rocks. Sexual

Genus *Petrogale*

(pet'-roh-gah'-lay: "rock-weasel")

The rock-wallabies of the genus *Petrogale* comprise a group of macropods adapted to life on rock-piles or rocky slopes. They share similar body shape and habits but differ considerably in coloration and size. Adaptations to locomotion among rocks include a short, broad foot with granulations over the entire sole, and a slender, barely tapering tail which is somewhat longer than the head and body. The tail is much more mobile than in other macropodids and is employed as a balancer during leaps, which can involve simultaneous forward, vertical and lateral components. Rock-wallabies may climb trees, particularly those with sloping trunks.

They are primarily grazers, but they may supplement this diet with leaves and fruits. Little is known of the biology of individual species. Females are usually larger than males. It seems that—probably because of the difficulty of moving over rocky country—a young animal does not follow its mother immediately after vacating the pouch, but remains in shelter among the rocks until it is completely weaned.

Short-eared Rock-wallaby

Petrogale brachyotis (brak'-ee-oh'-tis: "short-eared rock-weasel")

The Short-eared Rock-wallaby moves from the rocks into surrounding grassland to feed at night. During the day it sleeps among boulders or in and ferns. By day it sleeps in rock crevices. Breeding is probably continuous. One young is born.

HABITAT: rugged sandstone areas in tropical woodland
HEAD AND BODY: 30–36 cm
TAIL: 26–29 cm
DISTRIBUTION: 10,000–30,000 km²
ABUNDANCE: common
STATUS: probably secure

(I Morris)

Warabi

Petrogale burbidgei (ber'-bid-jee: "Burbidge's rock-weasel", after A. Burbidge, Australian zoologist)

Warabi is an Aboriginal name for this very small rock-wallaby (less than 1.5 kilograms). It feeds at night (also by day in the wet season) on grasses and ferns. By day it sleeps in a rock crevice. It probably breeds continuously with a peak in the summer wet season.

HABITAT: low rocky hills, cliffs and gorges in tropical savanna grassland
HEAD AND BODY: 41–52 cm
TAIL: 42–55 cm
DISTRIBUTION: 300,000–1 million km²
ABUNDANCE: abundant
STATUS: secure

(J Lochman)

81

Godman's Rock-wallaby

Petrogale godmani (god'-man-ee: "Godman's rock-weasel", after F. D. Godman, sponsor of T. V. Sherrin, who collected first specimen)

Godman's Rock-wallaby feeds at night on grasses, forbs and shrubs, often forming aggregations. It probably breeds continuously throughout the year.

HABITAT: tropical rainforest to dry sclerophyll forest with rocky slopes
HEAD AND BODY: 45–53 cm
TAIL: 48–59 cm
DISTRIBUTION: 100,000–300,000 km²
ABUNDANCE: common
STATUS: probably secure

(RL Close)

Unadorned Rock-wallaby

Petrogale inornata (in'-or-nah'-tah: "unadorned rock-weasel")

The name of the species refers to its having less brilliant colour or contrast than the other rock-wallabies. It inhabits some of the wettest areas in which rock-wallabies are found and it is less dependent than other species on the presence of rocks: it often climbs sloping trees. In the late afternoon and night it feeds on grasses.

Sexual maturity is reached at about 18 months. Breeding is continuous throughout the year, with a peak of births from March to July. One young is born. The species exhibits embryonic diapause.

HABITAT: steeply sloping rainforest to sclerophyll forest and woodland, often without much outcropping rock
HEAD AND BODY: 45–55 cm
TAIL: 49–60 cm
DISTRIBUTION: 100,000–300,000 km²
ABUNDANCE: common
STATUS: probably secure

(GA & MM Hoye)

Black-footed Rock-wallaby

Petrogale lateralis (lat'-er-ah'-lis: "side [-marked] rock-weasel", referring to pale stripe behind the shoulder)

The Black-footed Rock-wallaby feeds mainly on grasses in the late afternoon and evening, sometimes forming feeding aggregations. It may bask in sun during cool weather.

Sexual maturity is reached in the second year. One young is born. The species exhibits embryonic diapause.

HABITAT: tropical to temperate semiarid to arid granite rock-piles with mallee or other scrub cover, and adjacent grassland
HEAD AND BODY: 45–53 cm
TAIL: 48–60 cm
DISTRIBUTION: 300,000–1 million km²
ABUNDANCE: sparse
STATUS: probably secure

(D Mathews)

Brush-tailed Rock-wallaby

Petrogale penicillata (pen'-is-il-ah'-tah: "brush [-tailed] rock-weasel")

(RL Close)

The Brush-tailed Rock-wallaby is possibly the most sure-footed of all the rock-wallabies, being able to negotiate almost vertical rock faces. It moves into grassy areas to graze and browse on a wide variety of plants. By day it sleeps in a rock cleft or cave.

HABITAT: cliffs and rock slopes in subtropical to cool-temperate wet or dry sclerophyll forest with grassy understorey or adjacent grassy areas
HEAD AND BODY: 45–58 cm
TAIL: 52–67 cm
DISTRIBUTION: 100,000–300,000 km²
ABUNDANCE: common
STATUS: secure

Proserpine Rock-wallaby

Petrogale persephone (per-sef'-on-ay: "Proserpine rock-weasel", Persephone being the Greek name for the Roman goddess Proserpine; refers to town of Proserpine, Queensland)

The Proserpine Rock-wallaby, described in 1982, is limited to a very small area of rainforest. It grazes at night in open country and shelters by day among rocks.

Breeding is probably continuous through the year. One young is born. The species exhibits embryonic diapause.

HABITAT: rocky outcrops with surrounding woodland with grassy understorey, within tropical rainforest
HEAD AND BODY: 52–64 cm
TAIL: 60–68 cm
DISTRIBUTION: less than 10,000 km²
ABUNDANCE: rare
STATUS: possibly endangered

(J & D Bartlett)

Rothschild's Rock-wallaby

Petrogale rothschildi (roths'-chile-dee: "Rothschild's rock-weasel", after Lord Rothschild, sponsor of the expedition which collected the first specimens)

Rothschild's Rock-wallaby probably grazes and browses at night. It sleeps by day in rock clefts and caves.

HABITAT: subtropical to tropical shrubland and grassland with granite outcrops and rock-piles
HEAD AND BODY: 43–58 cm
TAIL: 47–67 cm
DISTRIBUTION: 100,000–300,000 km²
ABUNDANCE: common
STATUS: probably secure

(B & B Wells)

Yellow-footed Rock-wallaby

Petrogale xanthopus (ksan'-thoh-poos: "yellow-footed rock-weasel")

(U Schürer)

The Yellow-footed Rock-wallaby grazes at night on native grasses and browses on shrubs. It may also drink, but there are indications that it can survive for long periods without access to water; young have been seen to lick saliva from the mother's lips. Feeding aggregations are common. During the day, it sleeps among vegetation between boulders or in a rocky cleft.

Sexual maturity is reached between one and two years. Breeding is continuous throughout the year. The species exhibits embryonic diapause.

HABITAT: arid rock-piles and outcrops with open woodland and acacia scrubland; sometimes associated with sources of water
HEAD AND BODY: 48–65 cm

TAIL: 56–70 cm
DISTRIBUTION: 30,000–100,000 km²
ABUNDANCE: sparse
STATUS: vulnerable

Genus *Setonix*

(set-on'-ix: "bristle-claw")

The characteristics of this genus are those of the single species. It is distinguished by its short tail and hind feet.

Quokka

Setonix brachyurus (brak'-ee-ue'-rus: "short-tailed bristle-claw")

(J Lochman)

This small macropod has a relatively short tail and the claws of its short hind feet are almost obscured by long, stiff hairs. It was widespread through south-western Australia in the nineteenth century but, on the mainland, it now is restricted to small populations. It is abundant on Rottnest Island in dense semiarid heath, where it makes runways through the ground vegetation and feeds at night on native grasses and the leaves of shrubs. It requires access to drinking water. The Quokka is sociable and a number of animals may sleep together during the day in the shelter of dense vegetation.

Males are larger than females. In the more favourable conditions of the mainland, breeding is continuous but on Rottnest Island mating is restricted to January, February and March. A single young is born. The species exhibits embryonic diapause.

HABITAT: wet and dry sclerophyll forest, woodland and semiarid health
HEAD AND BODY: 40–55 cm
TAIL: 25–31 cm
DISTRIBUTION: 30,000–100,000 km²
ABUNDANCE: very sparse
STATUS: vulnerable

Genus *Thylogale*
(thie'-loh-gah'-lay: "pouched-weasel")

Pademelons are rather short-footed, compact-bodied macropods with tails that are relatively a little thicker and shorter than those of typical wallabies. Largely restricted to rainforest and wet sclerophyll forest, they combine browsing with grazing. Some individual pademelons have a vestigial upper canine tooth, reminiscent of the potoroid condition.

Tasmanian Pademelon

Thylogale billardierii (bil-ard'-ee-air'-ee-ee: "Billardière's pouched-weasel", after J. J. H. la Billardière, French naturalist who collected first specimens)

The Tasmanian Pademelon makes runways in dense ground vegetation. At night it moves to clearings on the forest edge to graze and eat the soft leaves of shrubs, often forming large aggregations. By day it sleeps deep in the undergrowth.

Males are much larger than females. Sexual maturity is reached in the second year. Breeding is continuous, with a peak of births from April to June. One young is born.

HABITAT: cool rainforest, wet sclerophyll forest and wet areas of dry sclerophyll forests with dense ground cover
HEAD AND BODY: 60–66 cm
TAIL: 30–48 cm
DISTRIBUTION: 30,000–100,000 km^2
ABUNDANCE: common
STATUS: secure

(D Greig)

Red-legged Pademelon

Thylogale stigmatica (stig-mat'-ik-ah: "pricked pouched-weasel", referring to faint [pricked out as with a pin on paper] markings on neck and hip)

In the northern part of its range the Red-legged Pademelon feeds mainly on the fallen leaves of rainforest trees; in the more southern parts, it includes more grass in its diet. Berries and the leaves of shrubs are also eaten. It sleeps from mid-morning to mid-afternoon with its back against a tree or rock, its tail tucked forward between its hind-legs, and its head on the tail.

Males are larger than females. It rears one young but we have no information on its breeding biology.

HABITAT: tropical rainforest to temperate wet sclerophyll forest
HEAD AND BODY: 39–54 cm
TAIL: 30–48 cm
DISTRIBUTION: 30,000–100,000 km^2
ABUNDANCE: common
STATUS: secure

(R & A Williams)

Red-necked Pademelon

Thylogale thetis (thay'-tis: "Thétis pouched-weasel", after French exploration vessel Thétis)

The Red-necked Pademelon makes runways in dense ground vegetation, moving out from these at night to graze in more open areas and to browse on herbs and soft-leaved shrubs. During the day it sleeps in dense vegetation. It is solitary but many form small breeding aggregations.

Males are markedly larger than females. Sexual maturity is reached at about 18 months. Mating takes place in January and February. One young is born.

(P Klapste)

HABITAT: temperate rainforest and wet sclerophyll forest, usually with adjacent grassland
HEAD AND BODY: 29–62 cm
TAIL: 27–51 cm
DISTRIBUTION: 100,000–300,000 km^2
ABUNDANCE: common
STATUS: probably secure

Genus Wallabia

(wol-ah'-bee-ah: "wallaby")

The characteristics of the genus *Wallabia* are those of the single species. In its browsing habit, skull anatomy, biochemistry, reproduction and chromosomes, it stands apart from the other members of the Macropodinae and its relationships are not yet understood.

Swamp Wallaby

Wallabia bicolor (bie'-col-or: "two-coloured wallaby")

The scientific name of this species refers to the contrast between dark brown upper parts and reddish orange underparts. The common name refers to its favoured habitat of swamps and damp gullies. It is solitary but may form feeding aggregations at night when it moves out from shelter to browse on shrubs and ferns, supplemented by grasses. During the day it shelters in dense forest vegetation. Males are slightly larger than females.

The relationships of this species with other macropodids are not clear. It is the largest of the browsing macropods; its chromosomes differ markedly from those of wallabies in the genus *Macropus*; and, while it exhibits embryonic diapause, it is unique in that the mating which gives rise to a quiescent blastocyst takes place about a week *before* the birth of an established foetus, not shortly after its birth. For these and other reasons related to skeletal features, the species is placed in a genus of its own.

(E Beaton)

Males are markedly larger than females. Breeding is continuous through the year. The single young remains in the pouch for eight to nine months but is suckled for another six to seven months at foot.

HABITAT: tropical to cool-temperate rainforest, sclerophyll forest and woodland with dense understorey
HEAD AND BODY: 67–85 cm
TAIL: 64–86 cm
DISTRIBUTION: 300,000–1 million km^2
ABUNDANCE: common
STATUS: secure

Subfamily STHENURINAE

(sthen'-ue-ree'-nee: "*Sthenurus*-subfamily")

The Sthenurinae differ from the Macropodinae in a wide range of anatomical features, of which two are easily recognised: the hind foot is relatively shorter, resembling that of potoroids, and the lower incisors bite against the tips of the upper incisors (not gliding against their inner surfaces, as in macropodines). Sthenurines were once represented by a wide range of species, all but one of which became extinct before the end of the Pleistocene. The surviving species, *Lagostrophus fasciatus*, is thus of considerable scientific interest and particularly worthy of special conservation measures. Due to an early lack of recognition of its uniqueness, *Lagostrophus fasciatus* is known as a hare-wallaby. This is confusing, since it has no close relationship with hare-wallabies of the genus *Lagorchestes*.

Genus *Lagostrophus*

(lah'-goh-stroh'-fus: "turning-hare")

The characteristics of the genus are those of the single species. *Lagostrophus* can be distinguished from *Lagorchestes* by its naked muzzle, very hairy feet (the hair covering the claws), and dark transverse bands on the back and rump.

Banded Hare-wallaby

Lagostrophus fasciatus (fas'-ee-ah'-tus: "banded turning-hare")

(B & B Wells)

late spring. One young is reared. The species exhibits embryonic diapause.

In the nineteenth century it occupied much of south-western Australia, but it is now restricted to Bernier and Dorre Islands, off the coast of Western Australia.

HABITAT: semiarid woodland and scrubland
HEAD AND BODY: 40–45 cm
TAIL: 35–40 cm
DISTRIBUTION: less than 10,000 km²
ABUNDANCE: common
STATUS: vulnerable

This small macropod is distinguishable by a pattern of dark transverse bands on the rump. It browses at night on the leaves of shrubs but also includes some grass in the diet; it does not need to drink. It is largely restricted to dense vegetation, through which it makes pathways. By day it sleeps in the shelter of the vegetation; it does not make a nest.

Females are slightly larger than males, which are territorial and aggressive. Sexual maturity is reached at less than one year of age but mating does not usually begin until the second year. Breeding takes place throughout the year except in

Order NOTORYCTEMORPHIA

(noh'-toh-rik-te-mor'-fee-ah: "*Notoryctes*-order")

The single species in this order is the only marsupial to spend its life underground. Adaptation to subterranean life must have involved a long history of evolution from more "ordinary" ancestors but we have no fossil history of these. Moreover, the anatomy of the Marsupial Mole is so specialised that it retains little evidence of its relationships with other marsupials. It has usually been regarded as more closely related to the Dasyuromorphia than to the Diprotodontia but current opinion leans slightly more towards the latter.

Its place among the marsupials remains uncertain but its uniqueness is recognised by allotting it to an order of its own.

Family NOTORYCTIDAE

(noh'-toh-rik'-tid-ee: "*Notoryctes*-family")

The characteristics of the family are those of the species.

Genus Notoryctes

(noh'-toh-rik'-tayz: "southern-digger")

The characteristics of the genus are those of the species.

Marsupial Mole

Notoryctes typhlops (tif'-lops: "blind southern-digger")

(*M Douglas*)

HABITAT: subtropical to tropical sandy deserts
HEAD AND BODY: 12–16 cm
TAIL: 2–3 cm
DISTRIBUTION: more than 1 million km²
ABUNDANCE: very sparse
STATUS: probably secure

In its adaptation to a burrowing life in desert sands, the Marsupial Mole has become blind and has lost its external ears. The blunt snout is protected by a horny shield, the tail is a tiny appendage, and the claws of the feet are broad digging blades. The body is covered with long, silky hair.

Because it spends almost all of its time below ground, it is seldom seen and little is known of its biology. Apparently it does not tunnel but, in effect, "swims" through the sand, leaving no cavity behind it. It has been assumed to feed on subterranean insect larvae and the few animals that have been held for short periods in captivity have survived on a diet of mealworms. However, one individual has been seen to eat a gecko, holding the body in a scissor-like grip between the claws of its forefeet. It does not need to drink.

The female has two teats in a well-formed pouch which opens backward. The male lacks a scrotum, the testes being within the abdomen.

(A supposed species, *N. caurinas*, was described in 1920 from Ninety Mile Beach, W.A. It is not considered to be sufficiently well-founded to be included here.)

Subclass EUTHERIA

(yue'-thee'-ree-ah: "perfect-mammals")

This large group comprises the more familiar mammals—those that are neither monotremes nor marsupials. They differ in many respects from these, but they could be defined as mammals which give birth to young with fully formed hind limbs. Females are characterised by having separate apertures for voiding faeces, urine, and young. Young are nurtured in the womb by means of a well-developed placenta.

Eutherian mammals have undergone a very diverse evolutionary radiation which includes burrowing moles, flying bats, and aquatic whales, seals and dugongs. There are many large and successful herbivores, and equally efficient carnivores which prey upon these.

The only eutherians native to Australia are bats (order Chiroptera), rodents (order Rodentia) and seals (order Carnivora). The Dingo (also a member of the Carnivora) is a "new Australian", resident for some thousands of years.

Order CHIROPTERA

(kie-rop'-te-rah: "hand-wings")

This order comprises the bats—mammals that fly by means of thin membranes of skin that extend between the fingers, to the hind limbs and often to the tail. In becoming specialised for flight, bats had to sacrifice some agility, but many of them can climb quite well, using one or two claws on the forelimbs and on the five toes. Some can even run (or scurry) on the ground or on the roof of a cave. When resting or sleeping, bats usually hang upside-down, suspended by their toes.

Almost anything that is said about bats has to be qualified by the statement that there are two very different kinds. Members of one group, typified by the flying foxes, feed mostly on fruit and/or nectar, are relatively large and find their way about by sight and scent; these comprise the suborder Megachiroptera and are often referred to as "megabats". Members of the other group, the Microchiroptera, are primarily insectivorous, are relatively small, and navigate mostly by ultrasonic echolocation: these are termed "microbats".

The differences between the two groups are so great that an increasing number of zoologists suspect that they may not be related through a common ancestor. It now seems possible that the similarity of the wings of microbats and megabats may be the result of the same sort of convergent evolution that has led eared seals and "true" seals to develop a similar body form. Indeed, some zoologists suggest that the megabats may be closely related to the primates, the mammalian group to which humans belong.

If the two groups of bats should prove to have evolved independently from non-flying ancestors, the order Chiroptera will have to be divided into two orders. While recognising this possibility, this book presents a conservative view.

Suborder MEGACHIROPTERA

(meg'-ah-kie-rop'-te-rah: "large-handwings")

Members of this group, the "megabats", have relatively large eyes and uncomplicated ears. The snout is usually rather long, the tail short or absent. The first and (usually) the second fingers bear claws. When roosting, a megabat hangs by the claws of its feet and wraps its wings around the body, the head being held at a right angle to the chest.

Most megabats feed by crushing fruit between their jaws, spitting out the pulp, and swallowing the juice and small seeds. Since the gut is short and simple, food passes through it in a few hours.

Some of the smaller megabats feed mainly on nectar. These also contribute to pollination of the flowers from which they obtain their food: many of these are large and open only at night.

Family PTEROPODIDAE

(te'-roh-poh'-did-ee: "*Pteropus*-family")

This is the only family of the Megachiroptera. It extends from tropical and subtropical Africa and Asia to the Australian region but a few species extend into southern Africa and south-eastern Australia. Feeding exclusively on fruits and nectar, pteropodids are largely restricted to rainforest and other forests which produce fruits or flowers throughout the year.

The family is divided into four subfamilies, three of which are represented in Australia: the nectar-feeding Macroglossinae and Nyctimeninae, and the fruit-eating Pteropodinae.

Subfamily MACROGLOSSINAE

(mak'-roh-glos-een'-ee: "*Macroglossus*-subfamily")

The two genera of blossom-bats which comprise this subfamily are characterised by possession of a long, flexible tongue with a brush-like tip: this is used to lap nectar. The nostrils are not tubular as in the Nyctimeninae. Two genera occur in Australia: *Macroglossus* and *Syconycteris*.

Genus Macroglossus

(mak'-roh-glos'-us: "long-tongue")

This genus includes three species of long-tongued, nectar-feeding bats which occur from India to the Philippines. They are all small and capable of hovering flight. The only one of these three species to extend its range to Australia is *M. minimus*.

Northern Blossom-bat

Macroglossus minimus (min'-im-us: "smallest long-tongue")

(*GA & MM Hoye*)

During the day it roosts, individually or in small groups, under large leaves in dense foliage. It flies slowly and can hover while sipping nectar.

Births are known to occur in August and September.

HABITAT: monsoon forest and tropical woodland, including bamboo thickets
HEAD AND BODY: 6–7 cm
TAIL: 1 cm
DISTRIBUTION: 300,000–1 million km^2
ABUNDANCE: common
STATUS: probably secure

This bat was recently regarded as a purely Australian species, *M. lagochilus*, but the Australian population now appears to be merely a racial variant of *M. minimus*, which extends through South-East Asia and Melanesia. It is similar in size, appearance and behaviour to the Queensland Blossom-bat *Syconycteris australis*, and distinguishable from it mainly by the large gap between its tiny lower incisors. It feeds with its long tongue on nectar and pollen (taking little, if any, fruit).

Genus *Syconycteris*

(sie'-koh-nik'-te-ris: "fig-bat")

This genus, which includes only two species, extends from the Moluccas to the Bismarck Archipelago. Like the related *Macroglossus*, it comprises small bats with long tongues that are used to feed on nectar. They can hover while feeding from a flower. Only one species of *Syconycteris* is found in Australia: *S. australis*.

Queensland Blossom-bat

Syconycteris australis (os-trah'-lis "southern fig-bat")

For a pteropodid, this is a very small bat (about 15 grams). It has a long, rather dog-like snout, similar to that of a flying-fox, and a long, protrusible tongue with which it collects nectar and pollen from a wide variety of forest and woodland trees. A solitary bat, it roosts in dense foliage. It flies below the canopy and is capable of hovering while it laps nectar and pollen from a blossom.

Mating takes place in October or November and a single young is born about 11 months later.

HABITAT: rainforest to sclerophyll forest and woodland
HEAD AND BODY: 4–6 cm
TAIL: vestigial
DISTRIBUTION: 100,000–300,000 km^2
ABUNDANCE: common
STATUS: probably secure

(*GA & MM Hoye*)

Subfamily NYCTIMENINAE

(nik'-tee-may-neen'-ee: "*Nyctimene*-subfamily")

Members of this subfamily feed on nectar by means of a protrusible tongue. A marked peculiarity of the nyctimenines are their tubular nostrils; the group is also unusual in having lower canine teeth which bite against the upper incisors (lower incisors are absent) and in having spotted wings and ears.

Only one genus, *Nyctimene*, occurs in Australia.

Genus *Nyctimene*

(nik'-tee-may'-nay: "night-moon")

These small nectar-feeding bats have protuberant tubular nostrils of unknown function. The snout is short and rounded and the wings and ears bear brightly coloured spots. Only the first digit of the forelimbs bears a claw. There is only one pair of incisors in the lower jaw. A tail is present.

Species of this genus are found from eastern Indonesia to the Solomons. They are known as tube-nosed bats.

Queensland Tube-nosed Bat

Nyctimene robinsoni (rob'-in-sun-ee: "Robinson's night-moon", after H. C. Robinson, who collected the first specimen)

This is the only species of the genus in Australia and is readily distinguished by its remarkably short, round face, tubular nostrils and yellow-spotted wings and ears. It feeds on flowers and fruits of trees and upon cultivated tropical fruits. It is not communal, roosting singly in dense foliage, changing the site as it moves nomadically in search of flowering or fruiting trees.

HABITAT: rainforest to wet and dry sclerophyll forest and woodland
HEAD AND BODY: 10–11 cm
TAIL: 2–3 cm
DISTRIBUTION: 100,000–300,000 km²
ABUNDANCE: common
STATUS: probably secure

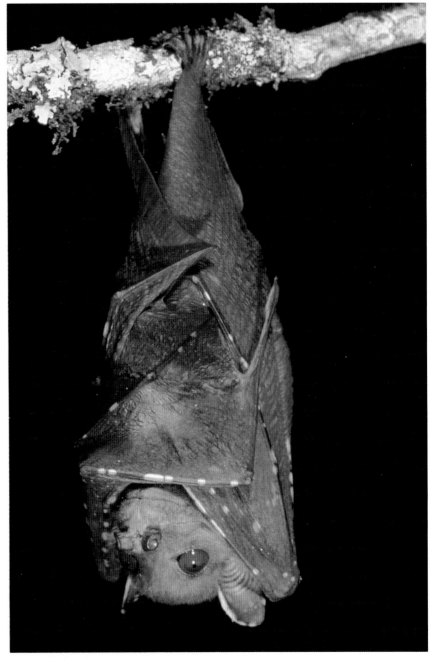

(R & A Williams)

Subfamily PTEROPODINAE

(te'-roh-pod-een'-ee: "*Pteropus*-subfamily")

This large group includes the fruit-bats and flying-foxes, characterised by possession of a simple tongue which cannot be extended far beyond the front of the mouth. Only two genera are represented in Australia: *Dobsonia* and *Pteropus*.

Genus Dobsonia

(dob-soh'-nee-ah: "Dobson's [bat]", after G. E. Dobson, British authority on bats)

Species of this genus, which extends from eastern Indonesia to the Solomons, have the wing membranes attached near the middle of the back, rather than to the sides of the body; there is thus a pocket-like space between the wings and the flanks. Unlike most other pteropodids, these bats do not have a claw on the second finger. A tail is present. There is only one pair of incisors on the lower jaw. They are often referred to as spinal-winged bats.

Bare-backed Fruit-bat

Dobsonia moluccense (mol'-uk-en'-say: "Dobson's [bat] from the Moluccas")

(H Millen)

The Australian population of the Bare-backed Fruit-bat is a southern outlier of a mainly New Guinean species and the only member of the genus in Australia. It is readily distinguished from other Australian bats by what appears to be a naked back, due to the wings meeting in the midline (with furred skin below the wings). The diet includes flowers and fruits of trees growing in the rainforest or near to its edges. It roosts by day in caves, rock crevices or very dense foliage. Colonies range from scores to hundreds of bats. The flight is slow, with rather fast flapping which permits great manoeuvrability, including hovering.

Mating takes place in May or June. A single young is born between September and November and carried by its mother for about a month, then left in the colony while the mother forages.

HABITAT: tropical rainforest
HEAD AND BODY: 28–32 cm
TAIL: 2–3 cm
DISTRIBUTION: 30,000–100,000 km²
ABUNDANCE: rare
STATUS: probably secure

Genus *Pteropus*

(te'-roh-poos: "wing-foot")

This is by far the largest genus of megabats, with more than 60 species. It is centred on southern Asia but extends westward to Madagascar and eastward to the islands of the South Pacific. It is characterised by a long, fox-like snout, simple nostrils and small, simple ears. The first two digits of the forelimbs bear claws. There is no tail. The lower jaw has two pairs of incisors.

Black Flying-fox

Pteropus alecto (ah-lek'-toh: "Fury wing-foot", referring to Alekto, one of the Furies of Greek mythology)

This species is more restricted to well-watered forest than the Grey-headed and Little Red Flying-foxes. It tends to roost in denser foliage and higher trees. It feeds on a wide variety of flowers and fruits of eucalypts and rainforest trees and may travel very long distances (up to 50 kilometres) from a camp to a feeding area. The Black Flying-fox is an agile climber. It has a fast and direct flight with rapid wing-beats, sometimes interspersed with short periods of gliding.

Pair-formation is followed by mating in March or April. The female gives birth to a single young from September to November and carries this with her for about a month. For about two months more it is left in the colony while the mother is foraging, being suckled daily on her return.

(BG Thomson)

HABITAT: tropical and subtropical rainforest, monsoon forest, wet sclerophyll forest and mangroves
HEAD AND BODY: 24–26 cm
TAIL: vestigial
DISTRIBUTION: 30,000–1 million km²
ABUNDANCE: abundant
STATUS: secure

Spectacled Flying-fox

Pteropus conspicillatus (kon-spis'-il-ah-tus: "spectacled wing-foot")

(V Serventy)

Even more dependent upon a wet environment than the Black Flying-fox, this species is restricted to areas of high rainfall in coastal and montane northern Queensland. It feeds on a wide variety of fruits and blossoms of rainforest trees and on nectar and blossoms of eucalypts. It is an agile climber and flies fast and directly with rapid wing-beats. It forms more or less permanent camps, some small, some containing tens of thousands of individuals. These may be moved a short distance when trees become so denuded by the bats' activities that they no longer provide sufficient shelter.

Pairing and mating take place from March to May and females give birth to a single young from October to December. Details of parental care are unknown but probably resemble those of other Australian flying-foxes.

HABITAT: tropical rainforest, wet sclerophyll forest, mangroves
HEAD AND BODY: 22–24 cm
TAIL: vestigial
DISTRIBUTION: 100,000–300,000 km²
ABUNDANCE: abundant
STATUS: secure

Grey-headed Flying-fox

Pteropus poliocephalus (poh'-lee-oh-sef'-al-us: "grey-headed wing-foot")

This is the largest of the Australian bats, with a wingspan up to 1.3 metres. It is gregarious and roosts during the day in colonies of tens of thousands of individuals, older animals on the perimeter acting as sentinels. At night, all but the suckling young fly out from the camp to feeding grounds, often many kilometres distant. The diet consists of the nectar and flowers of eucalypts and other native trees, and soft native fruits. Cultivated fruits are seldom eaten. It is an agile climber and flies fast and directly with steady wing-beats.

Mating takes place in March and April. Pregnant females form a separate maternity camp in September and each gives birth to a single young in October. This is carried by the mother until about five weeks old, then left in the camp for a further five weeks while the mother forages, suckling it on her return. Males and females form joint camps around January, establish pairs and eventually mate.

Many populations engage in annual migrations. This tends to be southward in the summer but some migrations seem not to fit this pattern.

HABITAT: tropical to temperate wet and dry sclerophyll forest and mangroves
HEAD AND BODY: 23–28 cm
TAIL: vestigial
DISTRIBUTION: 300,000–1 million km²
ABUNDANCE: abundant
STATUS: secure

(HJ Pollock)

Little Red Flying-fox

Pteropus scapulatus (skap'-ue-lah'-tus: "shoulder [-marked] wing-foot")

(GA & MM Hoye)

This flying-fox, smallest of the four Australian species, has by far the largest range. Unlike the others, it is found well inland of the Dividing Range in areas of low rainfall and sparse, rather low trees. Its range overlaps with those of the other three species and it frequently shares their camps. Its own camps may contain tens of thousands of individuals. At night, adults fly out from the camp, often over long distances, to feed on nectar, blossoms and fruits of a wide variety of trees, including eucalypts and melaleucas. Cultivated soft fruits, particularly tropical species, are often eaten.

Mating takes place from October to December, after which the females establish a separate maternity camp. Each female gives birth to a single young which is carried with the mother for about a month, then left in the camp for about one month more, being fed daily by the returning mother until it is weaned. Mixed camps of adult males and females re-form in October, establish pairs and mate. Maturity is not reached until the age of 18 months and subadults from the previous year's matings form separate juvenile camps on the outskirts of the adult camp.

The Little Red Flying-fox is an agile climber and flies fast and directly with deliberate wing-beats. It sometimes obtains drinking water by skimming the surface of a river or lake, then sucking the wet fur of its chest.

HABITAT: tropical to cool-temperate rainforest, wet and dry sclerophyll forest, mangroves, semiarid woodland

HEAD AND BODY: 19–24 cm

TAIL: vestigial

DISTRIBUTION: more than 1 million km^2

ABUNDANCE: abundant

STATUS: secure

Suborder MICROCHIROPTERA

(mike'-roh-kie-rop'-te-rah: "small-handwings")

Most bats belong to this large group, which accounts for about 20 per cent of the species of living mammals. Microbats are usually small and insectivorous, but a small group known as "false vampires", which may reach a length of abut 25 centimetres and a wingspan of up to 60 centimetres, are active predators on other bats and on terrestrial vertebrates: the Ghost Bat is the only Australian representative. The fishing bats catch small fishes and aquatic insects at the surface of the water by grasping them in their rake-like feet; the Large-footed Myotis is the only Australian member of this group. A few vampire bats, restricted to tropical America, feed on the blood of mammals. Tropical America is also the home of a number of microbats which have secondarily adopted a diet of fruit or nectar (there are no megabats in the Americas).

Microbats differ from megabats in many significant features. The thumb is the only digit on the hand to bear a claw; the eyes are small; the ears are usually large and often of complex shape; the snout is short and many carry an elaborate noseleaf. When a microbat is roosting, its wings are usually folded against the sides of the body and its head either hangs downwards or is held at a right angle to the back (facing backwards over the shoulder).

An outstanding feature of microbats is their ability to use ultrasonic echolocation for navigation and capture of prey. Pulses of ultrasound are broadcast through the mouth or nostrils; reflections from all solid objects in the vicinity are returned to the ears and the brain is able to interpret these to form a "picture" of what is ahead.

Echolocation is amazingly effective, permitting a microbat to fly at speed in total darkness, avoiding all obstacles and catching elusive insects as small as midges and mosquitos.

Five families of microbats are represented in Australia: the Emballonuridae (sheathtail-bats); Megadermatidae (false vampires); Molossidae (mastiff-bats); Rhinolophidae (horseshoe-bats); and the Vespertilionidae, a large and diverse group which has no inclusive common name. Research is proceeding so rapidly that some of the species described will soon be subdivided or amalgamated.

Family EMBALLONURIDAE

(em-bal'-on-yue'-rid-ee: "*Emballonura*-family", after a European genus of sheathtail-bats)

A diagnostic feature of this family, the sheathtail-bats, is that the tail appears to poke through the tail-membrane from below, as if there were a hole in the membrane: actually, the tail is connected to the membrane by very flexible skin. Because of this arrangement, the tail-membrane can slide along the tail when the hind legs are brought forward, thus making it easier for the bat to use its legs for walking or for running over the wall of a cave. Sheathtail-bats have a rather long, pointed snout without a noseleaf. The ear has a tragus. The wings are narrow and so long that, when the bat is at rest, the tips are folded back over the rest of the membrane.

The family is represented on all continents. Only two genera, *Saccolaimus* and *Taphozous*, are represented in Australia; other species of this genus occur from Africa, through Asia, to the Philippines.

Genus *Saccolaimus*

(sak-oh-lay-mus: "throat-pouch")

This genus of typical sheathtail-bats is named in reference to bare patches of sunken skin on the throat of some species. These appear to be glandular but their function is unknown.

Yellow-bellied Sheathtail-bat

Saccolaimus flaviventris (flah'-vee-vent'-ris: "yellow-bellied throat-pouch")

This species, largest of all the Australian sheathtail-bats, is extremely widespread: it is the only member of the family to be found in New South Wales and Victoria. In addition to these features, it is readily distinguished by the pale yellow or white fur on its undersurface.

It flies fast and directly, high above the forest canopy, feeding on flying insects. During the day it roosts in tree-holes and this habit has probably contributed to its extensive distribution. Populations in the southern part of the range do not hibernate but appear to migrate to warmer areas during the winter.

HABITAT: tropical to cool-temperate rainforest to woodland
HEAD AND BODY: 7–9 cm
TAIL: 2–3 cm
DISTRIBUTION: more than 1 million km²
ABUNDANCE: rare
STATUS: probably secure

(*GB Baker*)

Papuan Sheathtail-bat

Saccolaimus mixtus (mix'-tus: "intermediate throat-pouch")

This little-known species also occurs in New Guinea. It feeds on flying insects caught as it flies high over the forest canopy. It is known to roost in caves but it may possibly also use other sites.

HABITAT: tropical sclerophyll forest
HEAD AND BODY: 7–8 cm
TAIL: 2–3 cm
DISTRIBUTION: 10,000–30,000 km²
ABUNDANCE: rare
STATUS: probably secure

(BG Thomson)

Naked-rumped Sheathtail-bat

Saccolaimus saccolaimus (sak'-oh-lay'-mus: "throat-pouch throat-pouch")

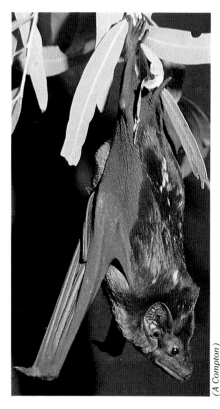

This species extends from India through South-East Asia and Indonesia to Papua New Guinea; the Australian populations are at the extreme south-eastern limits of the range.

The scientific name refers to a glandular pouch on the throat, a structure which is present in many sheathtail-bats and, when present, is usually better developed in males: its function is unknown. The common name refers to the absence of fur on the rump. It feeds on flying insects, taken as it flies fast above the forest canopy. During the day it nests, usually individually, in a hollow tree, house eaves or similar artefacts.

HABITAT: tropical rainforest, sclerophyll forest and woodland
HEAD AND BODY: 9–10 cm
TAIL: 2–4 cm
DISTRIBUTION: 300,0000–1 million km²
ABUNDANCE: rare
STATUS: probably secure

(A Compton)

Genus *Taphozous*
(taf'-oh-zoh'-us: "tomb-dweller")

This generic name was first applied to a species of sheathtail-bat described by Napoleon's scientific expedition to Egypt, from specimens found roosting in ancient tombs. Other species are now known from Africa and from South-East Asia to the Philippines.

North-eastern Sheathtail-bat
Taphozous australis (os-trah'-lis: "southern tomb-dweller")

This species, which also occurs in New Guinea, is similar to the Common Sheathtail-bat but smaller. It appears to be similar in feeding behaviour, preying upon flying insects such as beetles. During the day it roosts singly or in small numbers in small caves, rock fissures or the spaces in rock-piles.

It seems likely that mating occurs around April and that a single young is born in October or November.

(GA & MM Hoye)

HABITAT: tropical rainforest to woodland
HEAD AND BODY: 8–9 cm
TAIL: 2–3 cm
DISTRIBUTION: 100,000–300,000 km^2
ABUNDANCE: very sparse
STATUS: probably secure

Common Sheathtail-bat
Taphozous georgianus (jor'-jee-ah'-nus: "Georgian tomb-dweller", from King Sound, Western Australia)

(D Matthews)

The Common Sheathtail-bat has a very wide distribution in the semiarid regions of tropical and subtropical Australia. It feeds on flying insects, particularly beetles, which it obtains by patrolling a feeding area in a systematic manner. Its flight is direct and not very fast but it cannot hover like a horseshoe-bat. During the day, it roosts singly or in small numbers in caves, rock crevices or abandoned mines.

A single young is born in late spring or summer.

HABITAT: tropical and subtropical sclerophyll forest and woodland.
HEAD AND BODY: 6–8 cm
TAIL: about 3 cm
DISTRIBUTION: more than 1 million km^2
ABUNDANCE: common
STATUS: secure

Hill's Sheathtail-bat

Taphozous hilli (hill'-ee: "Hill's tomb-dweller", after J. E. Hill, British authority on bats)

(GB Baker)

Sheathtail-bat and no such structure in males of the Common Sheathtail-Bat. Hill's Sheathtail-bat flies fast and directly around trees in search of flying insects. It roosts by day in small colonies in caves or mines.

HABITAT: warm-temperate to subtropical arid woodland with caves or rock crevices.
HEAD AND BODY: 6–8 cm
TAIL: 2–4 cm
DISTRIBUTION: more than 1 million km²
ABUNDANCE: sparse
STATUS: probably secure

Like the Common Sheathtail-bat, Hill's Sheathtail-bat has a distribution which extends into desert regions. In fact, their distributions overlap and the two species, which are about the same size, often share the same roosts in caves or mines. They are difficult to distinguish but there is a glandular neck-pouch in the males of Hill's

White-striped Sheathtail-bat

Taphozous kapalgensis (kap'-al-gen'-sis: "Kapalga tomb-dweller", from place in Northern Territory)

One of the smaller sheathtail-bats, this species is distinguished by its pale brown or orange-brown fur with white stripes along the flanks. It flies high and fast over the canopy or low over water, presumably feeding on flying insects. It habits are not known but it is said to roost under the leaves of pandanus plants.

HABITAT: tropical monsoon forest and woodland
HEAD AND BODY: 7–9 cm
TAIL: about 2 cm
DISTRIBUTION: 10,000–30,000 km²
ABUNDANCE: rare
STATUS: probably secure

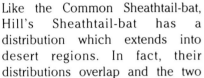

(D Matthews)

Family MEGADERMATIDAE

(meg'-ah-der-mat'-id-ee: "*Megaderma*-family")

Members of this family are the largest of the microbats, and Australia's Ghost Bat is the largest of these. False vampire-bats have very long, erect ears that are partially joined together over the forehead. The eyes are moderately large (*very* large for a microchiropteran) and there is an elaborate noseleaf. Both vision and echolocation are used in navigation and predation. These bats are carnivorous with a diet that includes other bats, birds, terrestrial mammals and reptiles as well as large insects.

The family is represented in Africa, India, South-East Asia, Melanesia, Australia and the Philippines.

Genus Macroderma

(mak'-roh-der'-mah: "great skin", in reference to the wings)

This genus, with only one species, is restricted to Australia. Its characteristics are essentially those of the species.

Ghost Bat

Macroderma gigas (jee'-gas: "giant great-skin")

The common name refers to the silent flight, pale colour and rather frightening appearance of this bat. It hangs from a tree or rock ledge, flying out to catch prey in the air or on the ground, then holding it in its wing while delivering killing bites. The dead animal is eaten at one of several feeding sites. Flight is fast, direct and powerful and a Ghost Bat has no difficulty in taking off from the ground, even when carrying prey. During the day it roosts in caves, in groups varying from 10 or so to about a hundred individuals.

Mating occurs in July or August and a single young is born from September to November. Pregnant and nursing females live in maternity colonies until the young are weaned. Young are not carried by the mother.

HABITAT: rainforest and arid woodland
HEAD AND BODY: 10–13 cm
TAIL: vestigial
DISTRIBUTION: more than 1 million km²
ABUNDANCE: very sparse
STATUS: vulnerable

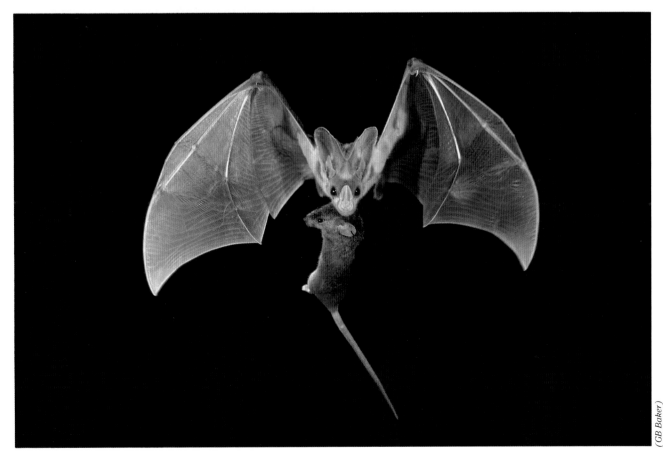

(GB Baker)

Family MOLOSSIDAE

(mol-os'-id-ee: "*Molossus*-family", after a genus of mastiff-bats from southern and central America)

Members of this family are known as mastiff-bats or free-tailed bats. The alternative common names refer to very visible features: the short, square muzzle with wrinkled lips has some resemblance to that of a mastiff; and the tip of the tail projects beyond the trailing edge of the tail-membrane (which, as in sheath-tailed bats, can move backwards and forwards along the tail). The wings are long and narrow, providing rapid and direct flight. The hind legs are short but well developed and have unusually strong musculature, permitting mastiff-bats to run with agility—hence they are sometimes referred to as "scurrying bats". Molossids lack a noseleaf and have large ears.

The family is represented on all continents in the tropical and warm-temperate regions of the Northern Hemisphere and extends into cool-temperate parts of the Southern Hemisphere. The Australian species belong to three very closely related genera, *Chaerophon*, *Mormopterus* and *Nyctinomus*.

Genus Chaerophon

(kee'-roh-fon: "pig-murderer")

Members of this genus of mastiff-bats differ from other Australian molossids in the following combination of characters. The upper jaw overhangs the lower jaw to a greater extent; the upper lip is very wrinkled; the inner margins of the ears are joined at their bases; and there is no pouch on the throat.

Northern Mastiff-bat

Chaerophon jobensis (joh-ben'-sis: "Jobi [Island] pig-murderer")

This chocolate-coloured bat is similar in many respects to the White-striped Mastiff-bat and replaces it in the northern third of the continent. The Australian population is a southern outlier of a species that is widespread through Melanesia.

It flies rapidly and directly over the canopy, feeding upon flying insects, and probably also takes some prey on the ground. It roosts in tree-holes, hollow trees, caves and human habitations, sometimes in hundreds. A single young is born in summer.

HABITAT: tropical sclerophyll forest to woodland
HEAD AND BODY: 8–9 cm
TAIL: 3–5 cm
DISTRIBUTION: more than 1 million km^2
ABUNDANCE: sparse
STATUS: probably secure

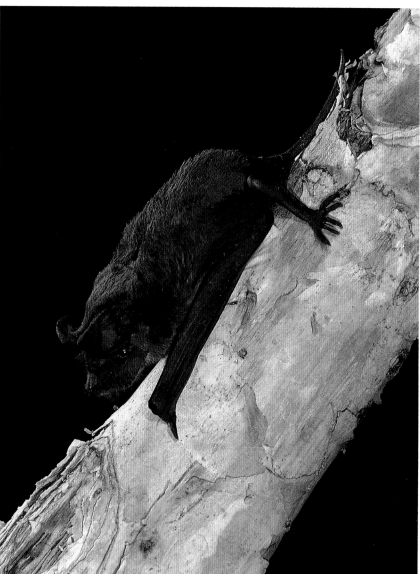

(BG Thomson)

Genus Mormopterus
(mor-mop'-te-rus: "monster-wing", from Mormo, a Greek mythical monster)

Members of this genus differ from other Australian mastiff-bats in the following combination of characters: the face is much less hairy; the upper lip is less wrinkled; the rather pointed ears are not joined at their inner margins; and there usually is no pouch on the throat (present in males of *M. norfolkensis*). The tail projects well beyond the tail-membrane.

Beccari's Mastiff-bat
Mormopterus beccarii (bek-ar'-ee-ee: "Beccari's monster-wing", after O. Beccari, Italian zoologist)

The common name of this species refers to its discoverer (possibly because it lacks distinguishing features that could be used in a descriptive name). In size, it is in the middle range of Australian molossids, overlapping with the Little Northern and Eastern Little Mastiff-bats. The Australian population is an outlier of a Melanesian species.

It flies fast above the canopy or over open water, feeding on a wide variety of flying insects. A variety of roosts, including tree hollows, spaces under peeling bark, and roof cavities are utilised. A single young is born, probably in summer.

HABITAT: subtropical and tropical rainforest to woodland
HEAD AND BODY: 5–7 cm
TAIL: 2–4 cm
DISTRIBUTION: more than 1 million km²
ABUNDANCE: abundant
STATUS: secure

(D Griffiths, L Hall)

Eastern Little Mastiff-bat
Mormopterus norfolkensis (nor'-fohk-en'-sis: "Norfolk [Island] monster-wing", mistakenly named after Norfolk Island)

Little can be said with certainty about the biology of this species, which is very similar to the Little Northern Mastiff-bat. However, it does not occur on Norfolk Island, being apparently restricted to subtropical and warm-temperate areas east of the Dividing Range. It appears to feed and roost in the same manner as the Little Northern Mastiff-bat.

HABITAT: warm-temperate to subtropical sclerophyll forest and woodland
HEAD AND BODY: 4–6 cm
TAIL: 3–4 cm
DISTRIBUTION: 30,000–100,000 km²
ABUNDANCE: very sparse
STATUS: probably secure

(GB Baker)

Little Mastiff-bat

Mormopterus planiceps (plah'-nee-seps: "flat-headed monster-wing")

The Little Mastiff-bat occupies much of the range of the White-striped Mastiff-bat, but has somewhat greater preference for arid habitats. It includes populations which, until recently, were regarded as a separate

(D Matthews)

species, the little Northern Mastiff-bat, *M. loriae*. It is much smaller (about one-third the weight) than the White-striped Mastiff-bat and its relatively flattened head permits it to roost in very narrow cavities or crevices, such as tree-holes, spaces under peeling bark and narrow spaces in human habitations.

This apparent species actually comprises two species, one of which has not yet been formally described. They can be distinguished by the length of the penis, the species with the larger penis having the more southern distribution.

Like other mastiff-bats, it feeds mainly on flying insects, which it takes in rapid and direct flight above the canopy, but it also chases crawling insects on the ground or on the trunks of trees.

A single young is born in summer.

HABITAT: dry sclerophyll forest to arid woodland
HEAD AND BODY: 5–7 cm
TAIL: 3–4 cm
DISTRIBUTION: more than 1 million km²
ABUNDANCE: common
STATUS: secure

Genus *Nyctinomus*

(nik'-tee-noh'-mus: "night-wanderer")

Members of this genus differ from other Australian molossids in the following combination of characters: the muzzle is very wrinkled; the inner margins of the ears are close together but not joined; and there is a well-developed pouch on the throat.

White-striped Mastiff-bat

Nyctinomus australis (os-trah'-lis: "southern night-wanderer")

This species is characterised by a white stripe along the flanks, just below the wings, and by rather fleshy ears. It occupies an unusually wide range of habitats, perhaps made possible by its very adaptable roosting habits: it spends the day in

hollow trees, tree-holes, under the peeling bark of trees and in human habitations, usually in small numbers. Flying fast and directly above the canopy, it feeds on flying insects but it may descend to the ground to chase and pounce upon

crawling insects.

Little is known of its reproduction but it appears that a single young is born around December, possibly from a mating 11 months previously.

HABITAT: temperate to subtropical sclerophyll forest to arid woodland
HEAD AND BODY: 8–10 cm
TAIL: 4–6 cm
DISTRIBUTION: more than 1 million km²
ABUNDANCE: common
STATUS: secure

(N Speechley)

Family RHINOLOPHIDAE

(rie'-noh-loh'-fid-ee: "*Rhinolophus*-family")

This family has some 70 species spread over most of the tropical and temperate parts of the world except the Americas. The noseleaf is noticeably longer than broad.

The common name for members of this family, horseshoe-bats, refers to the noseleaf, which usually includes a horseshoe-shaped membrane surrounding the nostrils. Horseshoe-bats emit their ultrasonic calls through the nostrils and the elaborate noseleaf probably serves as a reflector. The ears are large and simple in shape. The wings are broad and provide great manoeuvrability; the legs are weak.

Some authorities divide this group into two families, Rhinolophidae (horseshoe-bats) and Hipposideridae (so-called leaf-nosed bats).

Genus Hipposideros

(hip'-oh-sid-air'-os: "horse-iron", hence horseshoe)

Species of this genus have a somewhat squarer noseleaf than those in the genus *Rhinolophus*). They are usually brown to reddish brown and some individuals are a bright orange. The toes have only two joints (three in *Rhinolophus*. They are sometimes referred to as leaf-nosed bats, sometimes as horseshoe-bats.

Dusky Horseshoe-bat

Hipposideros ater (ah'-ter: "black horse-iron")

The Dusky Horseshoe-bat usually has mottled greyish fur and dark wings. It feeds on small to medium-sized insects in the understorey of relatively open forest and woodland, flying slowly and with great manoeuvrability. During the day, it roosts in small humid caves or rocky crevices, hanging by its feet and separate from other individuals. These bats tend to congregate in roosts during the summer and to disperse during winter.

Mating occurs in April. A single young is born around November.

HABITAT: monsoon forest and vine-thickets and adjacent sclerophyll forest

HEAD AND BODY: 4–5 cm

TAIL: 2–3 cm

DISTRIBUTION: 300,000–1 million km^2

ABUNDANCE: very sparse

STATUS: secure

(GB Baker)

Fawn Horseshoe-bat

Hipposideros cervinus (ser-vee'-nus: "fawn-coloured horse-iron")

This species is common from South-East Asia to the Philippines; the Australian population is an outlier, referred to the subspecies *H. c. cervinus*. Despite its name it is not always fawn-coloured. Some individuals are grey, some grey-brown with a reddish tinge, and a few are bright orange. It feeds on medium-sized flying insects which it takes in the understorey of the forest or over water. Its flight is so manoeuvrable that it is able to hover and then dart upon passing prey. During the day it roosts, often singly, in a cave, hanging from the roof and well separated from other individuals.

The single young is born around November and is carried for some time by the mother.

HABITAT: tropical rainforest and vine-thickets
HEAD AND BODY: 5–6 cm
TAIL: 2–3 cm
DISTRIBUTION: 10,000–30,000 km²
ABUNDANCE: very sparse
STATUS: probably secure

(L Hall, P Helman, SK Churchill)

Diadem Horseshoe-bat

Hipposideros diadema (die'-ah-dem'-ah: "diadem horse-iron")

The Australian population comprises two subspecies, *H. d. inornatus* from the Northern Territory and *H. d. reginae* from northern Queensland, which are outliers of a species that extends from South-East Asia to the Philippines. It is three to five times the weight of other Australian horseshoe-bats and is also distinguishable by the scattered white patches in its fur. It feeds on large flying insects, either patrolling an area or hanging from the branch of a tree, ready to swoop on passing prey. Its flight is more direct and less fluttering than that of other Australian horseshoe-bats.

During the day it roosts, often singly, in a cave, rock crevice, hollow tree or equivalent artefact.

Females form maternity colonies. A single young is born around November.

(I Morris)

HABITAT: tropical rainforest and monsoon forest
HEAD AND BODY: 7–9 cm
TAIL: 3–4 cm
DISTRIBUTION: 300,000–1 million km²
ABUNDANCE: very sparse
STATUS: probably secure

Greater Wart-nosed Horseshoe-bat

Hipposideros semoni (se-moh'-nee: "Semon's horse-iron", after R. W. Semon, German zoologist)

The common name of this bat refers to a "warty" protuberance between and above the nostrils. It feeds on small to medium-sized flying insects near the floor of the rainforest or among shrubs in open forests and woodland. During the day it roosts individually in caves, rock crevices or equivalent artefacts.

(L Hall, K Helman, SK Churchill)

HABITAT: tropical rainforest to woodland
HEAD AND BODY: 4–5 cm
TAIL: 1–2 cm
DISTRIBUTION: 30,000–100,000 km²
ABUNDANCE: rare
STATUS: probably secure

Lesser Wart-nosed horseshoe-bat

Hipposideros stenotis (sten-oh'-tis: "narrow-eared horse-iron")

This species is slightly smaller than the Greater Wart-nosed Horseshoe-bat and the "wart" on the noseleaf is much shorter. It feeds on small to medium-sized flying insects taken in the forest understorey or among shrubs in woodland. Little is known of its biology but it has been found roosting in caves, rock crevices and mines. Its flight is slow and manoeuvrable in foliage and marked by sudden darts upon the prey.

(L Hall, P Helman, SK Churchill)

HABITAT: tropical sclerophyll forest and woodland
HEAD AND BODY: 4–5 cm
TAIL: 2–3 cm
DISTRIBUTION: 300,000–1 million km²
ABUNDANCE: very sparse
STATUS: probably secure

Genus *Rhinolophus*

(rie'-noh-loh'-fus: "nose-crest")

This genus of horseshoe-bats has a more complex noseleaf than is found in *Hipposideros*, with a vertical structure, the "lancet", situated above the "horseshoe" element. Females have a teat-like structure (false teat) in each groin, on which the single young bites firmly when being carried by its mother.

Eastern Horseshoe-bat

Rhinolophus megaphyllus (meg'-ah-fil'-us: "great-leaf nose-crest")

This common species has an extensive distribution along the eastern coast of Australia. It feeds on

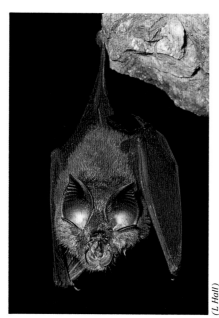

(L. Hall)

relatively large insects taken mostly in the understorey of forests. Its flight is slow and manoeuvrable among foliage and it is able to hover. During the day it roosts in a cave, hanging by its feet from the ceiling, seldom in contact with other individuals. In tropical regions colonies may consist of thousands of bats; in the southern part of the range, a cave may contain only a few. The Eastern Horseshoe-bat is active throughout the year in tropical habitats but may hibernate from April to June in cool-temperate habitats.

Mating usually occurs from April to June. Pregnant females segregate into maternity colonies. A single young is born around November and is carried by the mother for about a month, then left in the cave while the mother forages each day; it is weaned at the age of about six weeks.

HABITAT: tropical to cool-temperate wet and dry sclerophyll forest
HEAD AND BODY: 4–6 cm
TAIL: about 4 cm
DISTRIBUTION: 300,000–1 million km²
ABUNDANCE: common
STATUS: secure

Large-eared Horseshoe-bat

Rhinolophus philippinensis (fil'-ip-in-en'-sis: "Philippines nose-crest")

The Large-eared Horseshoe-bat has an extensive distribution from Sulawesi to the Philippines. The Australian population is a southern outlier, referred to the subspecies *R. p. robertsi*. It is characterised by having the most elaborate and "three-dimensional" noseleaf of all the Australian horseshoe-bats. It feeds on relatively large flying insects taken in the understorey of forest or over open water. The flight is slow and very manoeuvrable and it can hover and then dart at its prey.

Nothing is known of its reproduction in Australia.

HABITAT: tropical rainforest, vine-thickets and adjacent sclerophyll forest
HEAD AND BODY: 6–7 cm
TAIL: 3–4 cm
DISTRIBUTION: 100,000–300,000 km²
ABUNDANCE: rare
STATUS: probably secure

(L. Hall, P Helman, SK Churchill)

Genus Rhinonicteris
(rie'-no-nik'-te-ris: "nose-bat")

This genus is restricted to Australia. Its characteristics are those of the single species.

Orange Horseshoe-bat
Rhinonicteris aurantius (or-ant'-ee-us: "golden nose-bat")

Many species of horseshoe-bats have a small proportion of orange-coloured individuals, but all members of this species are orange. It is a small bat with a large, complex noseleaf and small ears, which feeds on small flying insects taken in the understorey of forest or around shrubs in woodland. During the day it roosts in deep caves, in colonies that vary in number from tens to thousands. Like other horseshoe-bats, it hangs by its feet from the roof at some distance from other individuals.

(GB Baker)

HABITAT: sclerophyll forest, woodland
HEAD AND BODY: 4–6 cm
TAIL: 2–3 cm
DISTRIBUTION: 300,000–1 million km^2
ABUNDANCE: very sparse
STATUS: probably secure

Family VESPERTILIONIDAE
(ves'-per-til-ee-on'-id-ee: "*Vespertilio*-family", from *Vespertilio*, a genus of small European bats)

About half of the Australian bat species are members of this family which, worldwide, has more than 300 species. Among the families of mammals it is exceeded in species only by the Muridae. In view of the size of the family and the fact that it is spread over most of the land surface of the earth, it is not surprising that it includes a wide variety of different types, making it difficult to define the family except in terms of rather subtle anatomical features. With few exceptions, however, vespertilionids are small and have minute eyes, simple ears which do not meet in the midline, and no noseleaf. The hind legs are not very strongly developed and the tail extends to the trailing edge of the tail-membrane or beyond it. They are mostly insectivorous, catching insects in flight, but some feed on the ground and a few catch fishes or aquatic insects with the claws of their hind feet.

Most species roost in caves but, across the family, just about every possible type of shelter, including human habitations, is utilised. The success of the vespertilionids in occupyintg cool-temperate and subpolar environments is related to the ability of many species to become torpid when food is scarce for relatively short periods or to hibernate for months during winter.

The family Vespertilionidae is classified into six subfamilies and, rather surprisingly in view of its isolation, the Australian bat fauna includes representatives of five of these.

Subfamily KERIVOULINAE
(ke'-ree-vue'-lin-ee: "*Kerivoula*-subfamily", after a genus of woolly bats)

Members of this small subfamily have woolly fur and funnel-like ears. Only one species occurs in Australia.

Genus *Kerivoula*

(ke'-ree-voo'-lah: from Kehevoulha, Sri Lankan name for a member of this genus)

Members of this genus have a rather long muzzle and a notably domed cranium, in reference to which they are sometimes called dome-headed bats. The upper canine teeth are long and dagger-like and have a groove along the outer side, in reference to which *Phoniscus* species are sometimes known as groove-toothed bats. The genus extends from South-East Asia through Melanesia to the Philippines.

Golden-tipped Bat

Kerivoula papuensis (pah'-pue-en-sis: "Papuan *Kerivoula*")

This small bat, the only Australian member of the subfamily Kerivoulinae, is readily recognised by its dark brown woolly fur with golden tips to the hairs. It occurs in New Guinea and has been found in widely distant locations on the eastern coast of Australia from Cape York to the vicinity of Eden. It flies fast or slowly and can hover as it gleans insects from leaves and branches below, or in, the canopy of rainforest. It appears to roost in dense vegetation and tree-holes.

HABITAT: cool-temperate to tropical rainforests
HEAD AND BODY: 5–6 cm
TAIL: 4–5 cm
DISTRIBUTION: 10,000–30,000 km^2
ABUNDANCE: very rare
STATUS: vulnerable

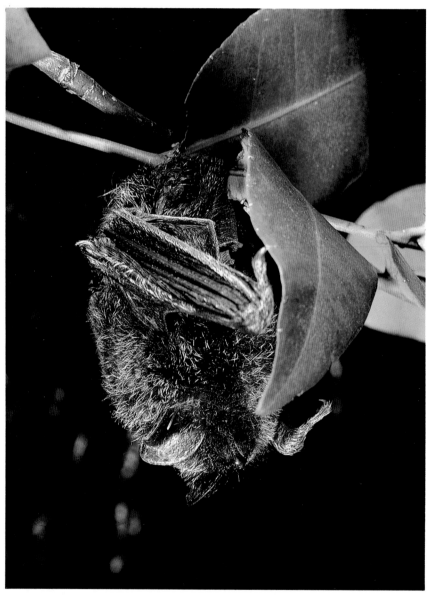

(GA & MM Hoye)

Subfamily MINIOPTERINAE

(min'-ee-op'-te-reen'-ee: *Miniopterus*-subfamily)

Members of this group have a very long third finger and a narrow wingtip.

Genus *Miniopterus*

(min'-ee-op'-te-rus: "small-wing")

This genus is distinguished by the flexure of the third digit under the upper part of the wing when the wing is folded. It is in reference to this that species of the genus are known as bent-wing bats. The tail is very long and there is a very obvious angle between the cranium and the snout.

Little Bent-wing Bat

Miniopterus australis (os-trah'-lis: "southern small-wing")

This species is similar to the Common Bent-wing Bat but significantly smaller. Like its larger relative, it roosts in caves, mines and houses and forages above the canopy, but its flight is slower.

Mating occurs around August. Pregnant females congregate in nursery caves.

HABITAT: tropical rainforest to warm-temperate wet and dry sclerophyll forest
HEAD AND BODY: 4–5 cm
TAIL: 4–5 cm
DISTRIBUTION: 300,000–1 million km²
ABUNDANCE: abundant
STATUS: secure

(A Young, D Gleason)

Common Bent-wing Bat

Miniopterus schreibersii (shrie'-ber-zee-ee, "Schreibers's small-wing", after K. F. A. von Schreibers, Austrian zoologist)

This species occurs almost throughout the world. It roosts communally in caves, mines and houses and forages at night above the forest canopy for small insects, flying fast and directly. It roosts communally, often in large numbers, in caves, mines and similar situations. Bats in the southern part of the range may hibernate in winter.

Mating occurs in May and June and in the southern part of the range but in September in the tropics. Pregnant females congregate in nursery caves.

HABITAT: tropical to cool-temperate wet and dry sclerophyll forest
HEAD AND BODY: 5–6 cm
TAIL: 5–6 cm
DISTRIBUTION: 300,000–1 million km²
ABUNDANCE: abundant
STATUS: secure

(GD Anderson)

111

Subfamily MURININAE

(myue'-ri-nee'-nee: "*Murina*-subfamily")

This group, which includes about a dozen species, extends from India and Japan to the Bismarck Archipelago. Its members are characterised by the possession of tubular nostrils.

Genus Murina

(mue-ree'-nah: "mouse-like")

This genus of small vespertilionid bats is characterised by tubular nostrils. Species occur from the Indian subcontinent through southern Asia, Indonesia and Melanesia to the Philippines. There is only one report from Australia.

Tube-nosed Insectivorous Bat

Murina florium (flor'-ee-um: "flower-associated mouse-like")

The nostrils of this bat are at the end of tubular structures that project sideways from the tip of the muzzle. It flies very slowly below the canopy, catching or gleaning small insects. During the day it roosts in the shelter of dense foliage. Its roosting posture is unique among Australian bats: the wings are curled around the body but out of contact with it, while the tail-membrane, curled over between the legs, overlaps the edges of the wings. The whole arrangement provides an umbrella.

The species is centred on New Guinea and there is only one record from Australia.

HABITAT: tropical misty rainforests
HEAD AND BODY: about 5 cm
TAIL: about 3 cm
DISTRIBUTION: less than 10,000 km^2
ABUNDANCE: very rare
STATUS: possibly endangered

(CA & MM Hoye)

Subfamily NYCTOPHILINAE

(nik'-toh-fil'-een-ee: "*Nyctophilus*-subfamily")

Nyctophilines can be distinguished from other members of the family Vespertilionidae by their possession of a small noseleaf.

Genus Nyctophilus

(nik'-toh-fil'-us: "night-lover")

In this genus the ears are large and usually connected by a low vertical membrane arising from the top of the head. The snout is short and bears a small, horseshoe-shaped noseleaf, seldom more than a ridge. They are known as long-eared bats.

Arnhem Land Long-eared Bat

Nyctophilus arnhemensis (ar'-nem-en'-sis: "Arnhem [Land] night-lover")

This species, which is smaller than the North Queensland Long-eared Bat, has a distribution which falls within the western range of that species. However, it appears to occupy a wetter local environment.

It has the slow, manoeuvrable flight typical of the genus and feeds on insects among or on foliage. It roosts in tree-holes, under bark, and sometimes in human constructions.

One or two young are born from October to February.

HABITAT: tropical monsoon forest and woodland, usually close to water
HEAD AND BODY: 4–5 cm
TAIL: 3–4 cm
DISTRIBUTION: 300,000–1 million km²
ABUNDANCE: common
STATUS: probably secure

(GB Baker)

North Queensland Long-eared Bat

Nyctophilus bifax (bie'-fax: "two-faced night-lover")

Little is known of the biology of this medium-sized long-eared bat. Like other members of the genus, it has a very manoeuvrable flight and forages for insects among foliage: it is able to hover while gleaning insects from leaves and tree-trunks or close to the ground. It can alight and take off from the ground with ease. During the day it roosts in tree-holes and sometimes in human constructions.

Two young are born, usually in December or January.

HABITAT: tropical rainforest to dry sclerophyll forest and woodland
HEAD AND BODY: 4–6 cm
TAIL: 4–5 cm
DISTRIBUTION: more than 1 million km²
ABUNDANCE: common
STATUS: probably secure

(BG Thomson)

Lesser Long-eared Bat

Nyctophilus geoffroyi (zhe-froy'-ee: "Geoffroy's night-lover", after E. Geoffroy Saint-Hilaire, French zoologist)

This species is a little smaller than Gould's Long-eared Bat and is distinguished from other Australian members of the genus by its very distinct noseleaf. It is a deliberate flier which forages for insects among foliage and close to the ground. It is extremely adaptable and occurs over almost the whole of Australia, including Tasmania. It roosts by day in tree-holes, under bark, in caves and in human constructions.

Animals in the southern part of the range may become torpid for periods in the depth of winter but the species is not known to hibernate.

Pregnant females segregate into maternity colonies and give birth, usually between October and January. Two young are usually born.

HABITAT: all Australian environments except tropical and subtropical rainforest
HEAD AND BODY: 4–5 cm
TAIL: 3–4 cm
DISTRIBUTION: more than 1 million km²
ABUNDANCE: abundant
STATUS: secure

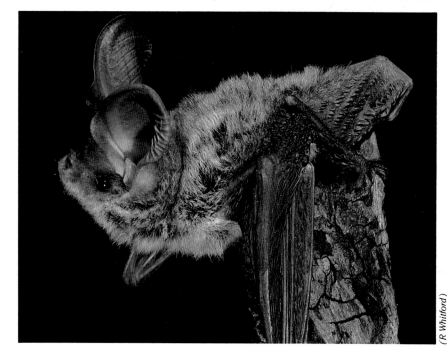

(R Whitford)

Gould's Long-eared Bat

Nyctophilus gouldi (gule'-dee: "Gould's night-lover")

This species resembles the Greater Long-eared Bat, to which it is closely related, but is smaller (about half the weight). Its range overlaps with that of the Greater Long-eared Bat but, in general, it inhabits wetter and more forested environments where the greater availability of food permits it to associate in large numbers. Small groups may roost together in tree-holes, under loose bark and in human constructions. It is a deliberate and manoeuvrable flier which captures insects flying among foliage and is even able to hover while gleaning insects from leaves. Animals in the southern part of the range hibernate during the colder parts of the year.

Two young are born in summer.

HABITAT: warm- to cool-temperate mallee and open woodland forest; subtropical rainforest
HEAD AND BODY: 5–7 cm
TAIL: 4–6 cm
DISTRIBUTION: 300,000–1 million km²
ABUNDANCE: common
STATUS: secure

(R Whitford)

Greater Long-eared Bat

Nyctophilus timoriensis (tee'-mor-ee-en'-sis: "Timor night-lover")

The Greater Long-eared Bat has a slow and deliberate flight and is even able to hover. It hunts large insects among the trees at the edges of rivers and creeks and over the water. Little is known of its biology but it is thought to roost in tree-holes or under loose bark.

It probably gives birth (perhaps to two young) from January to March.

HABITAT: semiarid to arid temperate woodland, particularly in vegetation fringing watercourses
HEAD AND BODY: 6–8 cm
TAIL: 4–5 cm
DISTRIBUTION: more than 1 million km²
ABUNDANCE: very sparse
STATUS: secure

(GB Baker)

Pygmy Long-eared Bat

Nyctophilus walkeri (waw'-ker-ee: "Walker's night-lover", after J. J. Walker, who collected first specimen)

This is the smallest Australian member of the genus, weighing about 4 grams. It is known from only a few specimens. Like the Arnhem Land Long-eared Bat, it is associated with water but, apparently, in rather drier local environments.

It seems that two young are normally born, probably in December or January.

HABITAT: tropical woodland fringing permanent water
HEAD AND BODY: 4–5 cm
TAIL: 3–4 cm
DISTRIBUTION: 30,000–100,000 km²
ABUNDANCE: very sparse
STATUS: probably secure

(GC Richards)

Subfamily VESPERTILIONINAE

(ves'-per-til'-ee-on'-een'-ee: "*Vespertilio*-subfamily")

This group includes those members of the Vespertilionidae which have virtually no distinguishing features. They can perhaps be regarded as the most "typical" of the microbats.

Genus Chalinolobus

(kal'-in-oh-lobe'-us: "bridle-lobe")

The name of this genus refers to lobes or wattles at the sides of the lower lip, roughly in the position of the bit in the bridle of a horse; the outer margin of the ear lies low down on the head, near the angle of the mouth. Some species have striking colour patterns. Members of the genus are usually referred to as wattled bats.

Large Pied Bat

Chalinolobus dwyeri (dwie'-er-ee: "Dwyer's bridle-lobe", after P. D. Dwyer, Australian zoologist)

This species is smaller than Gould's Wattled Bat and is characterised by woolly black fur and a white stripe along the flank, below the wing.

It flies rather slowly, feeding on insects below the canopy. During the day it roosts in a variety of locations including caves, old mines and tree-holes. Usually two young are born in November or December. These are able to fly when about three to four weeks old.

HABITAT: warm-temperate to subtropical dry sclerophyll forest and woodland
HEAD AND BODY: 4–5 cm
TAIL: 4–5 cm
DISTRIBUTION: 300,000–1 million km²
ABUNDANCE: common
STATUS: secure

(M Robinson)

Gould's Wattled Bat

Chalinolobus gouldii (gule'-dee-ee: "Gould's bridle-lobe", after J. Gould, English zoologist and artist)

(GD Anderson)

HABITAT: cool-temperate to tropical rainforest to arid woodland
HEAD AND BODY: 6–8 cm
TAIL: 4–5 cm
DISTRIBUTION: more than 1 million km²
ABUNDANCE: abundant
STATUS: secure

This species occupies an unusually wide range of habitats between Tasmania and Arnhem Land, such versatility being related to its ability to roost in almost any natural or artificial cavity. It feeds on a large variety of slow-flying insects, usually taken in manoeuvrable flight below the canopy. Some prey, such as caterpillars, are gleaned from bark or leaves. In the southern part of the range, Gould's Wattled Bat may become torpid for periods during the winter, but it has not been demonstrated to hibernate.

Mating occurs around May in the southern part of the range and sperm is stored by the female for fertilisation around July. Two young are usually born around November.

Chocolate Wattled Bat

Chalinolobus morio (mo'-ree-oh: "Moros's bridle-lobe", referring to Greek mythological Moros, son of Night)

Slightly smaller than the Large Pied Bat, this species is distinguishable by the uniform chocolate brown colour of its woolly fur. It flies moderately fast above and below the canopy, feeding on flying insects. It roosts in tree-holes, under bark and in artificial structures; on the Nullarbor Plain it roosts in caves. Bats in the southern part of the range hibernate, but for a shorter period than other vespertilionids at the same latitude.

In the southern part of the range, two young are born in December or January. In the subtropical parts of the range, births are in October.

HABITAT: cool- to warm-temperate sclerophyll forest to arid woodland
HEAD AND BODY: 5–6 cm
TAIL: 4–5 cm
DISTRIBUTION: more than 1 million km²
ABUNDANCE: common
STATUS: secure

(BG Thomson)

Hoary Bat

Chalinolobus nigrogriseus (nig'-roh-griz-ay'-us: "black-grey bridle-lobe")

This species is a little smaller than the Chocolate Wattled Bat and is characterised by white tips to its woolly dark grey fur. The Australian population is a subspecies of a New Guinean species.

It flies at moderate speed and with great manoeuvrability below the canopy or over shrubs, feeding on moths and other flying insects; it also gleans some crawling insects from the bark of trees or on the ground. Two young are born, probably in midsummer.

HABITAT: tropical to subtropical wet sclerophyll forest to woodland and heath
HEAD AND BODY: 4–6 cm
TAIL: 3–4 cm
DISTRIBUTION: more than 1 million km²
ABUNDANCE: very sparse
STATUS: secure

(BG Thomson)

Little Pied Bat

Chalinolobus picatus (pik-ah'-tus: "pied bridle-lobe", i.e. black and white)

This is the smallest Australian member of the genus *Chalinolobus*. It has similar markings to the Large Pied Bat but is only about half the weight. It is well adapted to aridity and to high temperatures. Its feeding behaviour has not been described but is probably similar to that of the Large Pied Bat. It roosts in caves, old mines, abandoned houses and possibly elsewhere. Two young are born around December.

HABITAT: warm-temperate to tropical semiarid to arid woodland
HEAD AND BODY: 4–5 cm
TAIL: 3–4 cm
DISTRIBUTION: more than 1 million km²
ABUNDANCE: sparse
STATUS: probably secure

(*GB Baker*)

Genus *Eptesicus*

(ep'-te-see'-kus: "house-flier")

Species of *Eptesicus* are small, undistinguished vespertilionids. They differ from pipistrelles in having only one premolar on each side of the upper jaw (pipistrelles have two). Species of the genus are found in all continents except the Americas. They are sometimes known as brown bats. The Australian species are all very similar and can seldom be identified with certainty except after examination of the skull and teeth. However, some species have distinct distributions.

Inland Eptesicus

Eptesicus baverstocki (bav'-er-stok'-ee: "Baverstock's house-flier", after P. Baverstock, Australian zoologist)

This species is almost indistinguishable from the Little Forest Eptesicus, *E. vulturnus* (within

which it was included until recently). It can only be differentiated from it on the anatomy of the skull and penis and by blood chemistry, but its distribution is quite distinct.

It is a fast and agile hunter of small insects below the tree canopy. It roosts in tree holes and in abandoned buildings.

HABITAT: semiarid to arid mulga and mallee woodland
HEAD AND BODY: 4–5 cm
TAIL: 3 cm
DISTRIBUTION: more than 1 million km²
ABUNDANCE: common
STATUS: secure

(*B & B Wells*)

Northwestern Eptesicus

Eptesicus caurinus (kor-ee'-nus: "northwestern house-flier", from Caurus, the northwest wind)

(BG Thomson)

This species, restricted to northwestern Australia, is not distinguishable on external characters from the other Australian eptesicuses but it is a little smaller than the Yellow-lipped Eptesicus (with which it shares part of its range) and it lacks the yellowish face of that species.

HABITAT: tropical woodland in deeply dissected rocky country
HEAD AND BODY LENGTH: 3–4 cm
TAIL LENGTH: 2–4 cm
DISTRIBUTION: 300,000–1,000,000 km²
ABUNDANCE: common
STATUS: probably secure

Large Forest Eptesicus

Eptesicus darlingtoni (dar'-ling-tun-ee: "Darlington's house-flier", after a Dr Darlington, who assisted in the capture of the first specimens)

Little is known of the biology of this bat, which (with a weight of about 8 grams) is the largest of the Australian eptesicuses. It feeds below the canopy on small insects, catching these with rapid changes of direction. It roosts in tree-holes and, occasionally, in buildings. It does not hibernate but, in the southern part of the range, becomes torpid for periods during the winter. In recent times the species has been known as *E. sagittula*.

The single young appears to be born around December.

HABITAT: cool-temperate wet and dry sclerophyll forest; elevated tropical woodland
HEAD AND BODY: 4–5 cm
TAIL: 3–4 cm
DISTRIBUTION: 100,000–300,000 km²
ABUNDANCE: common
STATUS: probably secure

Yellow-lipped Eptesicus

Eptesicus douglasorum (dug'-las-or'-um: "Douglas's house-flier", after A. Douglas, Australian zoologist and M. Douglas)

This species is similar in appearance and size to the Eastern Cave Eptesicus but is distinguished from it by its somewhat yellowish face. It roosts in caves and is known to catch at least some of its small insect prey over bodies of fresh water.

The single young appears to be born between August and December.

HABITAT: tropical monsoon forest and woodland
HEAD AND BODY: 3–4 cm
TAIL: 3–4 cm
DISTRIBUTION: 30,000–100,000 km²
ABUNDANCE: rare
STATUS: vulnerable

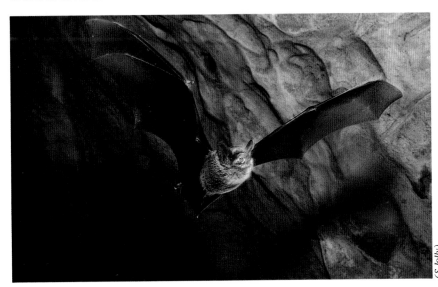

(S Jolly)

Western Cave Eptesicus

Eptesicus finlaysoni (fin'-lay-sun-ee: "Finlayson's house-flier", after H. H. Finlayson, Australian zoologist)

Like the related and slightly larger Eastern Cave Eptesicus, this species roosts in caves, rock fissures, mines and similar spaces in buildings. It may roost in pairs or in colonies of several hundred individuals. A fast and agile flier, it hunts below the tops of trees and often quite close to the ground.

It appears that breeding may occur any time of the year. Two young are normally born and reared.

HABITAT: tropical to warm temperate semiarid to arid woodland with appropriate subterranean shelter
HEAD AND BODY: 3–5 cm
TAIL: 3–4 cm
DISTRIBUTION: more than 1 million km²
ABUNDANCE: abundant
STATUS: secure

(*GD Anderson*)

Eastern Cave Eptesicus

Eptesicus pumilus (poom'-il-us: "dwarf house flier")

(*L. Hall*)

Territory, and from August to October in the Kimberleys. In the southern part of the range mating occurs in January and young are born about a year later, around December. Two young are usually born and these are carried for a time by the mother before being left in the roost.

HABITAT: warm-temperate to tropical woodland, sclerophyll forest and rainforest
HEAD AND BODY: 4–5 cm
TAIL: 3–4 cm
DISTRIBUTION: 300,000–1 million km²
ABUNDANCE: abundant
STATUS: secure

Like the Yellow-lipped Eptesicus and unlike other Australian members of the genus, this species roosts in caves. It feeds below the canopy on small flying insects which it catches by very agile, fluttering flight.

Until recently it was assumed to have had a distribution extending to the western Australian coast but *E. pumilus* is now restricted to the eastern coast, the other populations being assigned to a new species, *E. finlaysoni*.

The reproduction pattern varies with latitude. Breeding appears to be continuous in the northern part of the range, with peaks of births from March to June in the Northern

King River Eptesicus

Eptesicus regulus (reg'-ue-lus: "little-king house-flier", referring to King River, King George Sound, WA)

Little is known of the biology of this bat, which is largely restricted to

(*M Robinson*)

fairly well-watered temperate latitudes. It appears to roost in tree-holes and similar cavities.

Mating occurs around March and a single young is born around November.

HABITAT: cool-temperate sclerophyll forest to subarid woodland
HEAD AND BODY: 3–5 cm
TAIL: 3–4 cm
DISTRIBUTION: 300,000–1 million km²
ABUNDANCE: common
STATUS: probably secure

Troughton's Eptesicus

Eptesicus troughtoni (traw'-tun-ee: "Troughton's house-flier", after E. Troughton, Australian zoologist)

This recently described species, which cannot be identified on external characters, is largely restricted to northeastern Australia on both sides of the Dividing Range.

HABITAT: warm temperate to tropical woodland and sclerophyll forest
HEAD AND BODY LENGTH: 4–5 cm
TAIL: 3–4 cm
DISTRIBUTION: 300,000–1 million km²
ABUNDANCE: sparse
STATUS: probably secure

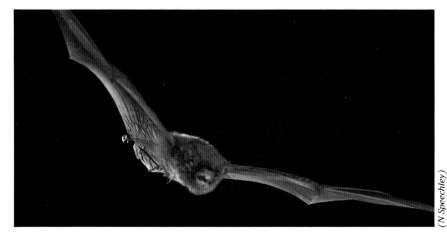

(*N Speechley*)

Little Forest Eptesicus

Eptesicus vulturnus (vul-ter'-nus: "vulture-like house-flier")

Despite its common name, this species extends well beyond the

(*R Whitford*)

limits of forests but, in general, it nests in tree-holes. Where these are unavailable, it utilises old buildings and other artefacts. It behaves like the other eptesicuses, flying below the canopy in a fluttering flight with rapid turns and pounces as it catches small flying insects.

In the southern part of the range individuals are largely inactive during the winter; they become torpid for periods but do not hibernate.

A single young is born around November or December while females are congregated into maternity colonies. Mating occurs shortly after giving birth.

HABITAT: cool-temperate to subtropical wet sclerophyll forest to arid woodland
HEAD AND BODY: 4–5 cm
TAIL: 3–4 cm
DISTRIBUTION: 300,000–1 million km²
ABUNDANCE: common
STATUS: secure

Genus *Falsistrellus*

(fol'-sis-trel'-us: "false-pipistrelle")

Only two species of this genus are known. The genus differs from *Pipistrellus* in having only one premolar on each side of the upper jaw (*Pipistrellus* has two).

Western Pipistrelle

Falsistrellus mckenziei (mk-ken'-zee-ee: "McKenzie's false-pipistrelle", after N. L. McKenzie, Australian zoologist)

This species is very similar in appearance to the Great Pipistrelle and has only recently been separated from it. The Western Pipistrelle probably has much the same behaviour as the Great Pipistrelle.

HEAD AND BODY LENGTH: 5–7 cm
TAIL LENGTH: 4–5 cm
DISTRIBUTION: 30,000–100,000 km^2
ABUNDANCE: sparse
STATUS: probably secure

HABITAT: cool temperate wet and dry sclerophyll forest

Great Pipistrelle

Falsistrellus tasmaniensis (tas-may'-nee-en'-sis: "Tasmanian false-pipistrelle)

Despite its scientific name, the Great Pipistrelle also occurs on the eastern coast of Australia and in south-western Western Australia. It is a large bat with a fluttering flight, which hunts larger insects below the forest canopy. It roosts in tree-holes, abandoned buildings and some-times caves. It appears that in the cooler part of the range some populations hibernate and others migrate in winter to warmer areas. A single young is born.

HABITAT: warm- to cool-temperate wet and dry sclerophyll forest
HEAD AND BODY: 5–7 cm
TAIL: 4–5 cm
DISTRIBUTION: 300,000–1 million km^2
ABUNDANCE: sparse
STATUS: probably secure

(*GA & MM Hoye*)

Genus *Myotis*

(mie-oh'-tis: "mouse-ear")

The name of this genus refers to the simple ears set well apart on the head and not connected to the corner of the mouth. It is a very large genus with species on all continents. They are sometimes referred to as mouse-eared bats or fishing bats.

Large-footed Myotis

Myotis adversus (ad-ver'-sus: "opposed mouse-ear", significance unknown)

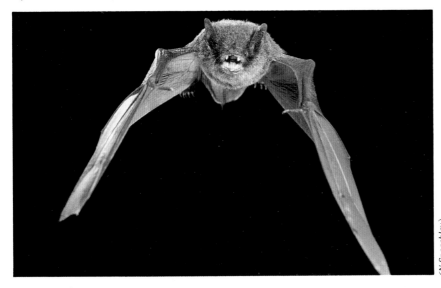

(*N.Speechley*)

This small bat (about 10 grams) feeds by skimming over the surface of placid water and grasping aquatic insects and small fishes with the rake-like toes and claws of its hind feet. Its range extends from eastern Indonesia through Melanesia, with an extensive coastal distribution in northern and eastern Australia. It roosts in caves, old mines and buildings, and in dense rainforest foliage. Populations in the southern part of the range undergo periods of torpor in the winter.

Breeding is unusually complicated. It is continuous in the tropics and three young may be born in a year. In the subtropics, one young is born around October and another around January. In the cool-temperate parts of the range, only one young is born, usually around December. Males roost separately from females except during the breeding season, when they associate in joint colonies based on harems, with each male dominating about a dozen females. Mating takes place shortly after the young are born.

HABITAT: cool-temperate to tropical wet sclerophyll forest
HEAD AND BODY: 5–6 cm
TAIL: 3–4 cm
DISTRIBUTION: 300,000–1 million km^2
ABUNDANCE: very sparse
STATUS: probably secure

Genus *Pipistrellus*

(pip'-is-trel'-us: "bat", from the Italian pipistrello, bat)

This genus has nearly 50 species spread over all continents. It is largely characterised by a *lack* of distinguishing features. The muzzle is broad and continues back to the head in a more or less straight line. There is no noseleaf. It differs from *Eptesicus* in having two upper premolars on each side (*Eptesicus* has one).

Northern Pipistrelle

Pipistrellus adamsi (ad'-am-zee: "Adams' bat", after M. A. Adams, Australian zoologist)

This small, delicate bat is difficult to distinguish from other Australian members of the genus.

Nothing in particular is known of its behaviour or reproduction.

HABITAT: tropical coastal forests
HEAD AND BODY LENGTH: 3–4 cm
TAIL: 3–4 cm
DISTRIBUTION: 100,000–300,000 km^2
ABUNDANCE: sparse
STATUS: probably secure

(*GA & MM Hoye*)

Northwestern Pipistrelle

Pipistrellus westralis (wes-trah'-lis: "Western Australian bat")

This species cannot be distinguished from Adams' Pipistrelle on external characters. Its behaviour is probably similar.

HABITAT: tropical coastal and near-coastal areas, including mangroves
HEAD AND BODY LENGTH: 3–4 cm
TAIL LENGTH: 3–4 cm
DISTRIBUTION: 100,000–300,000 km²
ABUNDANCE: sparse
STATUS: probably secure

Genus *Scoteanax*

(skoh'-tay-ah'-nax: "darkness-chief")

Closely related to *Scotorepens* and *Nycticeius*, this genus includes one or two species of broad-nosed bats, characterised by a wide and almost bare muzzle, relatively large eyes, and well-separated ears, the outer bases of which come near to the angles of the mouth.

Greater Broad-nosed Bat

Scoteanax rueppellii (rue-pel'-ee-ee: "Rüppell's darkness-chief", after W. P. E. S. Rüpell, German zoologist)

Largest of the Australian broad-nosed bats, this species flies slowly and directly in the understorey of forest and over water, feeding on large, slow-flying insects. It roosts in tree-holes and in buildings. A single young is born in January.

HABITAT: cool-temperate to tropical wet sclerophyll forest and rainforest
HEAD AND BODY: 8–10 cm
TAIL: 4–6 cm
DISTRIBUTION: 100,000–300,000 km²
ABUNDANCE: very sparse
STATUS: probably secure

(GA & MM Hoye)

Genus *Scotorepens*

(skoh'-toh-rep'-enz: "darkness-creeper")

Closely related to *Nycticeius* and *Scoteanax*, this genus resembles the latter except in minor aspects of skull anatomy. Species of *Scotorepens* are larger than those of *Scoteanax*.

Inland Broad-nosed Bat

Scotorepens balstoni (bawl'-stun-ee: "Balston's dark-creeper", after W. E. Balston, sponsor of expedition which discovered the species)

This is very similar to the Little Broad-nosed Bat and shares much of its range. However, it is not found to the east of the Dividing Range. It is a fast and agile predator of small flying insects, generally over or near to inland rivers or lakes. It usually drinks by skimming over the water surface. During the day it roosts in tree-holes or in similar artificial structures.

Mating occurs in May and one or two young are born in November. These are carried by the mother for about 10 days, after which they are left in the roost.

HABITAT: tropical to cool-temperate arid woodland, near permanent water
HEAD AND BODY: 5–7 cm
TAIL: 3–4 cm
DISTRIBUTION: more than 1 million km²
ABUNDANCE: common
STATUS: secure

(B & B Wells)

Little Broad-nosed Bat

Scotorepens greyii (gray'-ee-ee, "Grey's dark-creeper", after G. Grey, English explorer of Australia)

This species is much smaller than the Greater Broad-nosed Bat and is closely related to the Inland Broad-nosed Bat, with which its range overlaps considerably. It has a rather flattened head, which probably assists it to roost in narrow crevices in trees and buildings. It feeds on small flying insects.

Mating occurs around April or May. One or two young are born in November.

HABITAT: tropical to cool-temperate semiarid to arid woodland
HEAD AND BODY: 4–6 cm
TAIL: 2–4 cm
DISTRIBUTION: more than 1 million km²
ABUNDANCE: common
STATUS: secure

(R & A Williams)

Eastern Broad-nosed Bat

Scotorepens orion (o-rie'-on: meaning obscure, possibly implying "eastern")

Closely related to the Inland Broad-nosed Bat, this species differs in its preference for the wetter and cooler environments of coastal south-eastern Australia. It is a fast-flying bat which feeds, mainly below the canopy, on flying insects. It roosts in tree-holes and old buildings and intrudes into urban areas.

HABITAT: temperate to cool temperate wet and dry sclerophyll forest
HEAD AND BODY LENGTH: 4–5 cm
TAIL LENGTH: 3–4 cm
DISTRIBUTION: 100,000–300,000 km²
ABUNDANCE: sparse
STATUS: probably secure

Little Northern Broad-nosed Bat

Scotorepens sanborni (san'-born-ee: "Sanborn's dark-creeper", after C. S. Sanborn, American zoologist)

Smallest of the Australian broad-nosed bats, this species is found in areas of high rainfall on Cape York Peninsula, overlapping slightly with the Greater Broad-nosed Bat on the eastern coast and extending into New Guinea. It hunts small flying insects below the canopy, flying fast and with rapid changes of direction. During the day it roosts in a tree-hole or in a similar artificial crevice.

Mating probably occurs around May and one or two young are born around October.

HABITAT: tropical rainforest and monsoon forest to woodland, usually near permanent water
HEAD AND BODY: 4–5 cm
TAIL: 3–4 cm
DISTRIBUTION: 100,000–300,000 km²
ABUNDANCE: common
STATUS: probably secure

(*L Hall*)

Order RODENTIA

(roh-dent'-ee-ah: 'gnawers")

The most characteristic feature of rodents is the possession of one pair of large incisor teeth in the upper and in the lower jaws. The front surface of each incisor is covered with tough enamel but the rest of the tooth is composed of softer dentine and is worn away faster than the enamel, producing a chisel-like biting edge. Both the incisors and the grinding teeth grow continuously throughout the life of a rodent.

With more than 2000 species, rodents form the largest group of mammals. Great evolutionary radiation in Eurasia, Africa and the Americas has produced such animals as squirrels, flying-squirrels, gophers, jerboas, beavers, porcupines, agoutis, capybaras, rats, mice and voles but, of the two dozen families, only one, the Muridae (rats and mice), is represented in Australia.

The first rodents to reach Australia, probably about 10 to 15 million years ago, were members of the subfamily Hydromyinae. These make up the bulk of the Australian rodents, often referred to as the "old endemics". A second wave of immigrants brought in native rats of the genus *Rattus*, members of the subfamily Murinae, often referred to as the "new endemics". A third invasion, associated with European settlers, brought in the House Mouse and two other species of *Rattus*.

The subfamily Hydromyinae comprises three natural groupings, usually referred to as tribes. Such a fine degree of classification is not employed in respect of the other animal groups dealt with in this book but it is necessary in order to make some distinction among the old endemic rodents. The tribe Hydromyini, or water-rats, is a basically New Guinean assemblage with only two genera in Australia, *Hydromys* and *Xeromys*, each with one species. The Uromyini are also centred on New Guinea. One species of *Uromys* and five of *Melomys* are represented in Australia. The Conilurini are basically Australian, with only two species in New Guinea. These are the "real" old endemic rodents.

Suborder MYOMORPHA
(mie'-oh-mor'-fah: "mouse-form-suborder")

In addition to the typical rats and mice, this group includes such rodents as the flying (or gliding) squirrels, jerboas, jumping-mice, dormice, crested rats and bamboo rats. Of the five families, only one, the Muridae, occurs in Australia.

Family MURIDAE
(myue'-rid-ee: "mouse-family")

With at least 1200 species, this group includes more than a quarter of the world's living mammals. It is divided into 13 subfamilies but only two of these, the Hydromyinae and Murinae, are represented in Australia. No simple definition of the group can be given but most species are recognisably mouse-like or rat-like.

Subfamily HYDROMYINAE
(hie'-droh-mie-een'-ee: "*Hydromys*-subfamily")

It is impossible to define this subfamily on external characters and very difficult to do so on dental or skeletal anatomy. Much of its justification is based on evidence from chromosomes and blood proteins. However, in respect of the Australian hydromyines and the very limited number of native and introduced members of the subfamily Murinae, it is possible to make a general distinction. Except for the Prehensile-tailed Rat (*Pogonomys*), female hydromyines have only four teats. Australian members of the Murinae have eight to 12 teats.

Tribe CONILURINI
(kon'-il-yue-ree'-nee: "*Conilurus*-tribe")

Members of this tribe cannot be distinguished on external appearance. They are grouped together on the basis of rather subtle points of skull anatomy and on chemical similarities of their enzymes and other proteins. All evidence points to the Conilurini having been descendants of the first rodents to reach Australia. As the first colonisers, they had the time and opportunity to undergo the widest evolutionary radiation.

Genus Conilurus
(kon'-il-ue'-rus: "rabbit-tail")

Members of this genus are somewhat similar to those of *Mesembriomys* but have a blunter snout and protuberant eyes. The tail is brush-tipped. The fur is smooth. Rabbit-rats are arboreal. Females have four teats; the gestations period of the Brush-tailed Rabbit-rat is 33–36 days.

White-footed Rabbit-rat
Conilurus albipes (al'-bi-pez: "white-footed rabbit-tail")

This recently extinct species was mainly arboreal and nested in a tree-hole. Its diet is unknown. The female had four teats and reared up to three young which fastened themselves firmly to the mother's teats. It was last collected around 1840.

HABITAT: subtropical to cool-temperate wet and dry sclerophyll forest

HEAD AND BODY: 23–26 cm

TAIL: 22–24 cm

DISTRIBUTION: nil

ABUNDANCE: nil

STATUS: extinct

(J Gould)

Brush-tailed Rabbit-Rat

Conilurus penicillatus (pen'-is-il-ah'-tus: "brush [-tailed] rabbit-tail")

This species' brush-tipped tail is not diagnostic, being also characteristic of species of *Mesembriomys*. It is better recognised by its somewhat rabbit-like ears and head.

It is mainly arboreal, nesting in a tree-hole or similar shelter: it is also known to forage on the ground but details of its diet are not known.

The age at which sexual maturity is reached is not known. Breeding probably takes place throughout the year with peaks determined by rainfall. The female has four teats and normally rears one to three young which are independent at three weeks. Young attach themselves strongly to the mother's teats and are probably dragged behind her as she moves about.

(B & B Wells)

HABITAT: monsoon forest to tropical dry sclerophyll forest and pandanus scrub
HEAD AND BODY: 15–20 cm
TAIL: 18–22 cm
DISTRIBUTION: 300,000–1 million km^2
ABUNDANCE: sparse
STATUS: probably secure

Genus *Leggadina*

(leg'-ah-dee'-nah: "little-*Leggada*-like", *Leggada* being a genus of Indian mice)

The two species of *Leggadina* resemble species of *Pseudomys* but have smaller ears and a much shorter tail (not more than about 70 per cent of the head and body length). Short-tailed mice are terrestrial. Females have four teats. The gestation period is not known.

Forrest's Short-tailed Mouse

Leggadina forresti (fo'-res-tee: "Forrest's little-*Leggada*-like", after Sir John Forrest, explorer)

Hardly anything is known of the biology of this widespread and common mouse. It appears to be herbivorous, feeding on seeds and green vegetation. It probably does not need access to drinking water.

It shelters by day in a burrow.

It is known to breed after winter rains, but may be capable of breeding throughout the year. The female usually rears three or four young in a litter.

HABITAT: temperate to tropical arid tussock grassland
HEAD AND BODY: 8–10 cm
TAIL: 5–7 cm
DISTRIBUTION: more than 1 million km^2
ABUNDANCE: very sparse
STATUS: secure

(R Whitford)

Lakeland Downs Short-tailed Mouse

Leggadina lakedownenis (lake'-down-en'-sis: "Lake[land] Downs little-*Leggada*-like", from Lakeland Downs Station, northern Queensland)

Little is known of the biology of this species; its numbers may perhaps fluctuate greatly in response to the availability of grass seeds and green foods. It probably nests by day in a burrow but this has not been demonstrated.

It is probably capable of breeding throughout the year and of rearing a number of litters in rapid succession when food is abundant. Captive females have reared two to four young in a litter.

(H & J Beste)

HABITAT: not clearly established, but apparently includes tropical grassland, particularly where it borders on savanna
HEAD AND BODY: 6–8 cm
TAIL: 4–5 cm
DISTRIBUTION: 10,000–30,000 km^2
ABUNDANCE: rare
STATUS: probably secure

Genus Leporillus

(lep'-or-il'-us: "little-hare")

The most characteristic feature of the two species in this genus is that they nest inside piles of branches which are added to by successive generations. Stick-nest rats are large, with short, blunt heads, longish ears and eyes of moderate size. The fur is thick and soft and the tail is moderately hairy. They are terrestrial. Females have four teats. The gestation period of *L. conditor* is about 44 days.

Lesser Stick-nest Rat

Leporillus apicalis (ah'-pik-ah'-lis: "tipped little-hare", referring to the white-tipped tail)

Little is known of the biology of this recently extinct species (last collected in 1933) but it was probably similar to that of the Greater Stick-nest Rat.

HABITAT: arid woodland and shrubland
HEAD AND BODY: 17–20 cm
TAIL: 22–24 cm
DISTRIBUTION: nil
ABUNDANCE: nil
STATUS: extinct

(J Gould)

Greater Stick-nest Rat

Leporillus conditor (kon'-dit-or: "builder little-hare")

(HJ Aslin)

The Greater Stick-nest Rat feeds at night on succulent plants and some grasses. During the day it shelters in a nest within a large structure of branches and leaves (often exceeding 2 cubic metres) which has been maintained and added to by successive generations. Ten to 20 individuals may inhabit the stick-nest. Where appropriate sticks and branches are not available, they may dig burrows or utilise those made by other animals. The species has disappeared from the mainland and the only known population is on Franklin Island in the Nuyts Archipelago.

Mating probably occurs from March to June. The female usually rears one or two young which attach themselves to the teats and are dragged about by the mother. Young become independent at about four weeks.

HABITAT: warm-temperate to cool-temperate arid woodland or shrubland
HEAD AND BODY: 17–26 cm
TAIL: 14–18 cm
DISTRIBUTION: less than 10,000 km²
ABUNDANCE: very sparse
STATUS: endangered

Genus *Mastacomys*

(mas'-ta-koh-mis': "jawed-mouse")

This genus is restricted to Australia and has only one species. Its name refers to the very broad molars and powerful jaw muscles associated with the large component of grass and bark fibre in its diet. Females have four teats; the gestation period is about 35 days.

Broad-toothed Rat

Mastacomys fuscus (fus'-kus: "grey jawed-mouse")

(GB Baker)

This is a stockily built, broad-headed, short-tailed, long-furred rodent of a size equivalent to the lower range of Australian *Rattus* species. It is primarily a grass-eater and its broad molars and powerful jaw muscles are an adaptation to this difficult food. It does not burrow, but makes runways and builds a nest in dense ground vegetation which is often covered by snow in the winter. It is mainly nocturnal but may be active during the day in the colder parts of the year, particularly in Tasmania.

Sexual maturity is reached at about one year. Mating occurs from September to January in Tasmania and from November to January in the alpine country of New South Wales. The usual litter size is two; young cling firmly to the mother's teats for about three weeks, and are weaned at about five or six weeks. Two litters may be reared in a season.

HABITAT: cold, wet alpine and subalpine grassland with sedges and rushes, often near permanent water
HEAD AND BODY: 14–18 cm
TAIL: 10–13 cm
DISTRIBUTION: 30,000–100,000 km²
ABUNDANCE: sparse
STATUS: probably secure

Genus *Mesembriomys*

(mez-em'-bree-oh-mis: "southern-mouse")

The two species of this genus, which are confined to northern Australia, are almost as large as the Giant White-tailed Rat. The ears are large and the long tail is brush-tipped. The head is rat-like with large (but not protuberant) eyes. The fur is shaggy. Tree-rats are arboreal but may spend much time on the ground. Females have four teats; the gestation period of the Black-footed Tree-rat is 43 to 44 days.

Black-footed Tree-rat

Mesembriomys gouldii (gule'-dee-ee: "Gould's southern-mouse", after J. Gould, English zoologist)

(TJ Smith)

The Black-footed Tree-rat is an arboreal rodent which feeds on fruits, flowers and large seeds, supplemented by insects, snails and even mussels. It is one of the largest of Australian rodents and is readily recognised by its shaggy grey-brown hair. It seems to be mainly arboreal and nests in tree-holes, but it also spends some time on the ground, even entering the water to feed.

Males are a little larger than females. Sexual maturity is reached at about three months. Most breeding seems to occur from June to August. The female has four teats and rears one to three young which cling to the mother's teats and are dragged behind her as she moves about. The gestation period of a little more than six weeks is unusually long and the young, which are born in a more advanced state than other Australian rodents, are weaned at about six weeks.

HABITAT: tropical sclerophyll forest and woodland
HEAD AND BODY: 25–30 cm
TAIL: 32–41 cm
DISTRIBUTION: 300,000–1 million km²
ABUNDANCE: very sparse
STATUS: probably secure

Golden-backed Tree-rat

Mesembriomys macrurus (mak-rue'-rus: "long-tailed southern-mouse")

The Golden-backed Tree-rat is similar in general appearance to the

(C Kemper)

Black-footed Tree-rat but has smooth fur and is less than half the weight of the latter species. It is mainly arboreal, feeding on shoots, fruits and nuts and making a nest in a tree hollow. It also spends time on the ground and, in some areas, forages along the tide-line in search of oysters or stranded fishes. It will enter houses, where it eats food scraps, rice and flour.

The breeding season, if any, is not known. The female has four teats and usually rears one or two young.

HABITAT: tropical woodland, tussock grassland, pandanus scrub and vine-thickets
HEAD AND BODY: 19–25 cm
TAIL: 29–36 cm
DISTRIBUTION: 100,000–300,000 km²
ABUNDANCE: very sparse
STATUS: vulnerable

Genus *Notomys*

(noh'-toh-mis: "southern-mouse")

This genus is Australian in distribution, occurring mostly in arid country. Hopping-mice are characterised by long, slender hind legs which are employed in a hopping gait, and by a long brush-tipped tail which provides balance. The eyes and ears are large. Nests are in deep burrows. Many arid-adapted species do not need to drink, obtaining all their requirements from moisture in food and the water produced by metabolism of carbohydrates. Water conservation is aided by the choice of cool, humid nest-sites and by the production of highly concentrated urine and very dry faeces. Females have four teats. The usual gestation period appears to be around 38 to 41 days.

Spinifex Hopping-mouse

Notomys alexis (ah-lek'-sis: "Alexis southern-mouse", referring to Alexandria Downs Station, NT)

The Spinifex Hopping-mouse occupies most of the central desert areas of Australia, where it feeds on seeds, shoots, roots and insects. It obtains sufficient water from this diet (even from a diet of dry seeds) to have no need for drinking water; this economy is made possible by its extremely concentrated urine and very dry faeces. Water loss is also reduced by the habit of nesting during the day in groups of up to 10 individuals in deep, humid burrows.

Sexual maturity is reached at about two and a half months and breeding may occur at any time of the year: the actual times of breeding are probably determined by rainfall. The female has four teats and usually raises three or four young, which are weaned at four weeks. As a contribution to water conservation, the female drinks the urine produced by her suckling young.

HABITAT: arid cool-temperate to tropical spinifex hummock grass-land and sand, including desert dunes
HEAD AND BODY: 10–11 cm
TAIL: 13–15 cm
DISTRIBUTION: more than 1 million km²
ABUNDANCE: common
STATUS: secure

(D Matthews)

Short-tailed Hopping-mouse

Notomys amplus (am'-plus: "large southern-mouse")

This recently extinct species (not collected since 1895) was larger than any of the living hopping-mice (probably about twice the weight of *Notomys mitchelli*). It had large ears and a tail only slightly longer than its body. Nothing is known of its biology.

HABITAT: desert dunes or gibber plains
HEAD AND BODY: about 14 cm
TAIL: about 15 cm
DISTRIBUTION: nil
ABUNDANCE: nil
STATUS: extinct

Northern Hopping-mouse

Notomys aquilo (ak-wil'-oh: "northern southern-mouse")

Like the Dusky Hopping-mouse, this species is restricted to sand-dunes, but those inhabited by the Northern Hopping-mouse tend to be stabilised with acacia shrubs over ground cover of spinifex. Small groups occupy complex burrows. Nothing is known of the diet or reproduction.

HABITAT: tropical coastal dunes with significant vegetation
HEAD AND BODY: 9–11 cm
TAIL: 16–17 cm
DISTRIBUTION: 30,000–100,000 km²
ABUNDANCE: very sparse
STATUS: possibly endangered

Fawn Hopping-mouse

Notomys cervinus (ser-vee'-nus: "fawn southern-mouse")

(R Whitford)

The Fawn Hopping-mouse has much the same distribution as the Dusky Hopping-mouse and the two are similar in size. They are able to coexist because of strong differences in habitat preference: the Fawn Hopping-mouse prefers gibber plains, while the Dusky species is limited to dunes. The two may feed quite exclusively, even when only metres apart.

The Fawn Hopping-mouse feeds mainly on seeds, supplemented by green vegetation. It is able to drink salt water and excrete the excess salt in its urine. Its burrows are made in the excessively hard clay underlying gibber plains and it has been suggested that this would only be possible when the clay has been softened by rain.

Sexual maturity is reached at about six months. Breeding is possible throughout the year but is probably determined by rainfall. The female has four teats and usually rears three young which become independent at about four weeks.

HABITAT: arid gibber plains
HEAD AND BODY: 9–12 cm
TAIL: 10–16 cm
DISTRIBUTION: 30,000–100,000 km²
ABUNDANCE: sparse
STATUS: probably secure

133

Dusky Hopping-mouse

Notomys fuscus (fus'-kus: "dusky southern-mouse")

Like most members of its genus, the Dusky Hopping-mouse is adapted to desert life but it tends to be restricted to large sand-dunes, in which it makes extensive burrows communicating with the surface by a number of vertical shafts. It forages for seeds, grasses and some insects on the surface of the dunes but does not go into gibber areas, even when these are quite close. (Gibber areas are utilised by the Fawn Hopping-mouse.)

Sexual maturity is reached at two and a half to three months and, although breeding can occur at any time of year, it is probably determined mainly by rainfall. The female has four teats and commonly rears three young which are weaned at about four weeks.

HABITAT: desert dunes
HEAD AND BODY: 8–12 cm
TAIL: 11–16 cm
DISTRIBUTION: 10,000–30,000 km²
ABUNDANCE: rare
STATUS: endangered

(R Miller)

Long-tailed Hopping-mouse

Notomys longicaudatus (lon'-jee-kaw-dah'-tus: "long-tailed southern-mouse")

This recently extinct species (not collected since 1901) was a little larger than Mitchell's Hopping-mouse but had a relatively longer tail. Its burrow was made in dense soil or clay. Nothing is known of its biology.

HABITAT: arid to semiarid woodland, heath and hummock grassland, usually on clayey soil
HEAD AND BODY: about 13 cm
TAIL: about 18 cm
DISTRIBUTION: nil
ABUNDANCE: nil
STATUS: extinct

(J Gould)

Big-eared Hopping-mouse

Notomys macrotis (mak-roh'-tis: "long-eared southern-mouse")

Nothing is known of the biology of this recently extinct species, which shows some physical resemblance to the Fawn Hopping-mouse. It was last collected in 1844.

HABITAT: semiarid woodland
HEAD AND BODY: 11–12 cm
TAIL: 13–14 cm
DISTRIBUTION: nil
ABUNDANCE: nil
STATUS: extinct

Mitchell's Hopping-mouse

Notomys mitchelli (mit'-chel-ee: "Mitchell's southern-mouse", after Sir Thomas Mitchell, explorer and discoverer of species)

(R Whitford)

Largest of the hopping-mice, this species lives closer to centres of human population (Perth, Adelaide) than any of the others and is commonly exhibited in major Australian zoos. It feeds on seeds, green plants and some small insects: when these are scarce, it eats roots. Unlike the more desert-adapted hopping-mice, it requires regular access to drinking water. During the day, it sleeps in a nest in a deep burrow system communicating with the surface by a vertical shaft.

Females become sexually mature at about three months and are probably capable of breeding throughout the year. There are four teats and the usual litter is two to four young which are weaned at about four weeks and become independent about a week later.

HABITAT: cool- to warm-temperate semiarid woodland
HEAD AND BODY: 10–13 cm
TAIL: 14–16 cm
DISTRIBUTION: 30,000–100,000 km²
ABUNDANCE: sparse
STATUS: probably secure

Darling Downs Hopping-mouse

Notomys mordax (mor'-dax: "biting southern-mouse")

Nothing is known of the biology of this recently extinct species, which was described on the basis of a single skull, collected around 1844, supposedly from the Darling Downs.

HABITAT: woodland to open forest
HEAD AND BODY: insufficient data
TAIL: insufficient data
DISTRIBUTION: nil
ABUNDANCE: nil
STATUS: extinct

Genus Pseudomys

(sue'-doh-mis: "false-mouse")

This large genus includes a large number of species that superficially resemble the introduced House Mouse and rats. They cannot be distinguished, as a genus, on external features. They range in size from 10 grams (*P. delicatulus*) to nearly 120 grams (male *P. gracilicaudatus*).

Although it would be appropriate to call these animals pseudo-mice, they are generally referred to simply as mice: larger species are often known as rats, which is rather confusing, since Australia has a number of native and introduced species of the genus *Rattus* (true rats).

One (rarely two) species of *Pseudomys* can be found in virtually any part of mainland Australia, Tasmania and many offshore islands. One species occurs in New Guinea.

Habitats range from the most arid deserts to very wet southern rainforests, but most species are arid-adapted. Many of these require no access to drinking water, obtaining all their needs from the moisture in their food and from the water produced by the metabolism of carbohydrates. Water conservation is aided by seeking cool, humid nesting sites and by the production of very concentrated urine and dry faeces. ·

Species of *Pseudomys* are predominantly seed-eaters but a few feed mostly on grasses. Most supplement their diet with insects or other arthropods. They are all terrestrial.

Females have four teats. The gestation period of most species appears to be around 27 to 30 days but is 34 to 38 days in one of the smaller species (*P. delicatulus*) and 22 to 24 days in the slightly larger *P. nanus*.

Ash-grey Mouse

Pseudomys albocinereus (al'-boh-sin'-er-ay'-us: "ashy-white false-mouse")

Females of this species are slightly larger than the House Mouse, males markedly so. Both are much lighter in colour (ash-grey) than the House Mouse. During winter the Ash-grey Mouse feeds on seeds and green vegetation; in summer it eats a lot of insects. It probably does not need access to drinking water. During the day, individuals and family groups sleep in a nest in a complex burrow which may be 3 to 4 metres long.

Mating occurs from August to October on the mainland and perhaps from March to August on Bernier and Dorre Islands. Normally, only one litter is raised in a year. The female has four teats but litters of up to six young have been reared. Young become independent at six to seven weeks.

HABITAT: semiarid subtropical to cool-temperate woodland and heath on sandy soil
HEAD AND BODY: 6–10 cm
TAIL: 8–11 cm
DISTRIBUTION: 10,000–30,000 km²
ABUNDANCE: sparse
STATUS: probably secure

(AC Robinson)

Silky Mouse

Pseudomys apodemoides (ap'-oh-dem-oy'-dayz: "*Apodemus*-like false-mouse", *Apodemus* being a genus of European fieldmice)

The Silky Mouse is closely related to the Ash-grey Mouse and has a similar coloration but a softer coat. It is slightly smaller than a House Mouse and has much larger ears. It eats the seeds of casuarinas and leptospermums and, particularly in winter, it takes nectar from the Desert Banksia. Swarming cockroaches are also eaten. It shelters communally in a complex burrow extending as far as 3 metres below the surface and often constructed in the shade of a Desert Banksia. It invades areas after a bushfire and population densities reach a peak in regenerating vegetation two years after such a fire.

Breeding can occur at any time of the year, but appears to be strongly influenced by rainfall. Some populations breed in winter, others in spring and summer. Females may rear successive litters of three or four young.

HABITAT: cool-temperate semiarid mallee heathland, particularly around Desert Banksia shrubs
HEAD AND BODY: 6–9 cm
TAIL: 9–11 cm
DISTRIBUTION: 30,000–100,000 km²
ABUNDANCE: sparse
STATUS: probably secure

(IR McCann)

Plains Mouse

Pseudomys australis (os-trah'-lis: "southern false-mouse")

This species, which is about four times the weight of a House Mouse, is sometimes referred to as the "Plains Rat", but it is a typical mem-

ber of the genus *Pseudomys* and significantly smaller than most species of *Rattus*. It feeds mostly on seeds, supplemented with green plant material and some insects. It does not need to drink. During the day it shelters in a complex, shallow burrow system that extends over a large area. Non-breeding animals may congregate in groups of up to 20 or so; breeding animals stay in family groups. Populations fluctuate greatly according to rainfall.

Females become sexually mature at nine to 10 months. Breeding appears to be possible throughout the year but takes place mostly in winter and spring. The usual litter size is three or four young, which become independent at about four

(R Whitford)

weeks. Several litters may be reared in succession.

HABITAT: cool- to warm-temperate tussock or hummock grassland on gibber plans
HEAD AND BODY: 10–14 cm
TAIL: 8–12 cm
DISTRIBUTION: 300,000–1 million km²
ABUNDANCE: very sparse
STATUS: probably secure

Bolam's Mouse

Pseudomys bolami (boh'-lam-ee: "Bolam's false-mouse", after A. J. Bolam, Australian naturalist)

Until 1984, this was regarded as a subspecies of the Sandy Inland Mouse. Although it has been separated as a distinct species, we do not know how it differs in its biology.

HABITAT: temperate arid woodland with sparse scrub cover
HEAD AND BODY: 6–8 cm
TAIL: 8–10 cm
DISTRIBUTION: 30,000–100,000 km²
ABUNDANCE: sparse
STATUS: probably secure

Western Pebble-mound Mouse

Pseudomys chapmani (chap'-man-ee: "Chapman's false-mouse", after G. Chapman, Australian zoologist)

This species, which has a rather long head and large ears, is a little larger

than the Pilliga Mouse but significantly smaller than the House Mouse. Its most outstanding characteristic (shared with the Central Pebble-mound Mouse) is the construction, over successive generations, of large piles of pebbles, within which several individuals (or pairs) make tunnels and nests. The average weight of the pebbles is about 5 grams (half the average weight of an individual). The mounds provide a cool, humid shelter and represent an alternative to burrowing in very hard soil.

Nothing is known of its breeding pattern.

(B & B Wells)

HABITAT: arid tropical spinifex grassland and acacia woodland on pebbly soil
HEAD AND BODY: 5–7 cm
TAIL: 7–8 cm
DISTRIBUTION: 30,000–100,000 km²
ABUNDANCE: rare
STATUS: vulnerable

Delicate Mouse

Pseudomys delicatulus (del'-i-kat'-ue-lus: "rather delicate false-mouse")

This species is the smallest of the *Pseudomys* species, about half the weight of a House Mouse. It appears to feed mainly on grass seeds. During the day it sleeps in a nest made in a wide variety of shelters—in or under fallen timber or stones, or in a complex burrow which may be 2 metres long.

Males and females are sexually mature at 10 to 11 months. Mating occurs in June and July and the female normally rears two to four young which become independent at three to four weeks.

HABITAT: tropical grasslands, woodland, dry sclerophyll forest and coastal dunes
HEAD AND BODY: 5–8 cm
TAIL: 5–8 cm
DISTRIBUTION: more than 1 million km²
ABUNDANCE: sparse
STATUS: probably secure

(HJ Aslin)

Desert Mouse

Pseudomys desertor (dez-er'-tor: "desert false-mouse")

(BG Thomson)

HABITAT: arid hummock grassland and shrubland, often on dunes
HEAD AND BODY: 8–10 cm
TAIL: 8–10 cm
DISTRIBUTION: 30,000–100,000 km²
ABUNDANCE: very sparse
STATUS: probably secure

The Desert Mouse is a little larger than the House Mouse but has much longer fur, and larger ears and eyes (with a pronounced eye-ring). Little is known of its biology but in one study area it is known to feed on sedges and grasses. It does not need to drink. It is solitary, males and females occupying separate nests. These are made in shallow burrows or in dense vegetation, through which it also makes runways.

Sexual maturity is reached at about 10 weeks. Breeding can probably occur at any time of the year but is largely determined by rainfall. The usual litter is about three young, which become independent at about three weeks. Several litters may be reared in rapid succession.

Smoky Mouse

Pseudomys fumeus (fue-may'-us: "smoky false-mouse")

The Smoky Mouse is significantly larger than the House Mouse (two to three times heavier). Its food varies with the season. In summer it feeds mainly on seeds and berries; in winter mainly on underground fungi. In spring, it feeds to a considerable extent on migrating Bogong Moths. Populations are unstable, with a high mortality from September to November.

Breeding occurs in summer and one or two litters of three or four young may be reared.

(R Miller)

HABITAT: mainly subalpine sclerophyll forest and woodland with heath understorey, often on mountain ridges; also on coastal plains
HEAD AND BODY: 8–10 cm
TAIL: 11–15 cm
DISTRIBUTION: 10,000–30,000 km^2
ABUNDANCE: very sparse
STATUS: possibly endangered

Gould's Mouse

Pseudomys gouldii (gule'-dee-ee: "Gould's false-mouse")

This extinct species (last collected in 1857) was about twice the weight of a House Mouse. All we know of its biology is that it slept communally in burrows.

HABITAT: warm-temperate dry sclerophyll forest to semiarid woodland
HEAD AND BODY: 10–13 cm
TAIL: 9–10 cm
DISTRIBUTION: nil
ABUNDANCE: nil
STATUS: extinct

(J Gould)

Eastern Chestnut Mouse

Pseudomys gracilicaudatus (gras'-il-ih-kaw-dah'-tus: "slender-tailed false-mouse")

This close relative of the Western Chestnut Mouse is significantly larger: males are three to four times the weight of a House Mouse; females two to three times. It feeds largely on grasses, also on grass seeds. The nest is constructed above ground or in a burrow. Runways are made in the dense ground vegetation that is its normal habitat.

Mating occurs from about August to March. Females may rear three successive litters of about three young in rapid succession. Young are independent at about four weeks.

(R Whitford)

HABITAT: wet swampy heathland with dense shrubs, woodland with dense grass cover
HEAD AND BODY: 10–15 cm
TAIL: 8–12 cm
DISTRIBUTION: 300,000–1 million km^2
ABUNDANCE: very sparse
STATUS: probably secure

Sandy Inland Mouse

Pseudomys hermannsburgensis (her'-mans-berg-en'-sis: "Hermannsburg false-mouse", from Hermannsburg Mission, NT)

(R Miller)

HABITAT: very variable, extending over about two-thirds of Australia, from the arid tropics to arid cool-temperate regions, on sandy deserts, gibber flats, hummock grassland and mulga woodland
HEAD AND BODY: 6–9 cm
TAIL: 7–9 cm
DISTRIBUTION: more than 1 million km^2
ABUNDANCE: sparse
STATUS: secure

This species is a little larger than the Delicate Mouse but not quite as large as the House Mouse, which it otherwise resembles, although its eyes and ears are larger. It feeds on seeds, supplemented by green vegetation and tubers. It probably does not need to drink. During the day it sleeps—often in company with other individuals—in a nest at the end of a simple burrow up to about a metre long.

Breeding probably can occur at any time of the year but is likely to be stimulated by rainfall. The female can rear three or four young in a litter. These become independent at four to five weeks.

Long-tailed Mouse

Pseudomys higginsi (hig'-in-zee: "Higgins's false-mouse", after E. T. Higgins, British naturalist, who described the species [as *Mus leucopus*])

About three times the weight of the House Mouse, this species is readily distinguished by its slender tail, which is almost one and a half times the length of the head and body. It feeds on a wide range of green plant material, supplemented by insects and spiders. It seems that a male and female mate for life and, together with any dependent offspring, nest under fallen timber or in a shallow tunnel.

The Long-tailed Mouse is mainly nocturnal, but in cold weather it may be active during the day. In contrast to most *Pseudomys* species, which are adapted to hot, arid conditions and a fluctuating food supply, the Long-tailed Mouse lives in very wet, cold rainforest with a year-round assurance of food. It is probably because of this that permanent, sedentary, territorial pairs can be established.

Mating occurs from November to April. The female has four teats and normally rears three or four young in a litter which are independent at four to five weeks but remain with the parents until 12 to 13 weeks old. Suckling young are left in the (temporarily plugged) nest while the mother forages, but cling tightly to her teats if she is disturbed or moves into a new nest.

HABITAT: deeply littered floor of very wet Antarctic Beech rainforest
HEAD AND BODY: 11–15 cm
TAIL: 14–20 cm
DISTRIBUTION: 30,000–100,000 km²
ABUNDANCE: common
STATUS: probably secure

(JE Wapstra)

Central Pebble-mound Mouse

Pseudomys johnsoni (jon'-sun-ee: "Johnson's false-mouse" after K. A. Johnson, Australian zoologist)

This species was described in 1985 and little is known of its biology. It has not been directly observed to make pebble-mounds but it is assumed to have much the same behaviour as the Western Pebble-mound Mouse.

HABITAT: arid tropical acacia woodland and hummock grassland on stony ridges and plains
HEAD AND BODY: 7 cm
TAIL: *c.* 9 cm

DISTRIBUTION: less than 10,000 km²
ABUNDANCE: rare
STATUS: possibly endangered

Kimberley Mouse

Pseudomys laborifex (lab-or'-if-ex: "work-maker", referring to difficulty of finding specimens)

This mouse, which was described in 1986, is similar in appearance to several other species of *Pseudomys*. Among the similar species that share or approach its distribution, it is a little larger than the Delicate Mouse; it has brownish fur on the back rather than grey as in the Central Pebble-mound Mouse; and it is smaller than the Western Chestnut Mouse.

Nothing is known of its biology.

HABITAT: tropical woodland with spinifex or grass ground cover, sometimes rocky
HEAD AND BODY: 6–7 cm
TAIL: 7–9 cm
DISTRIBUTION: 30,000–100,000 km²
ABUNDANCE: very sparse
STATUS: probably secure

(Western Australian Museum & CM Kemper)

Western Chestnut Mouse

Pseudomys nanus (nah'-nus: "dwarf false-mouse")

This orange-fawn species is about twice the weight of the House Mouse and has much larger eyes, surrounded by a pale ring. It appears to feed on grasses and may not need to drink. Its nesting habits are unknown.

It probably breeds throughout the year, except in the dry period from September to November, rearing successive litters of about three young in rapid succession when conditions are most favourable. Young are independent at three weeks. The gestation period of 22 to 24 days is the shortest known in the hydromyine rodents.

HABITAT: dense tussock grassland, sometimes with cover of eucalypts
HEAD AND BODY: 8–14 cm
TAIL: 7–12 cm
DISTRIBUTION: 300,000–1 million km²
ABUNDANCE: sparse
STATUS: secure

(JA Kerle)

New Holland Mouse

Pseudomys novaehollandiae (noh'-vee-hol-an'-dee-ee: "New Hollandian [i.e. Australian] false-mouse"')

One of the most interesting features of the New Holland Mouse is that it was "rediscovered" near Sydney in 1967 after not having been recorded for about a century. Within a few years, it had been found in other parts of New South Wales, Victoria and even Tasmania. Clearly, we had not been looking well enough for it, or in the right places. It may often have been mistaken for the House Mouse, which is about the same size and has a very similar appearance.

The New Holland Mouse feeds mainly on seeds in the warmer part of the year and insects in the colder part, but also eats green vegetation and fungi. During the day, individuals and family groups sleep in a nest in tunnels up to five metres long.

Females become sexually mature at seven weeks or more but produce only one litter in the first year of life. Subsequently they may produce three or four litters in a year. Mating is from July to November and a female normally rears three or four young in a litter.

HABITAT: low heath on sandy soil
HEAD AND BODY: 6–9 cm
TAIL: 8–11 cm
DISTRIBUTION: 100,000–300,000 km²
ABUNDANCE: common
STATUS: secure

(R Whitford)

Western Mouse

Pseudomys occidentalis (ok'-sid-ent-ah'-lis: "western false-mouse")

Little is known of the biology of this greyish species, which is a little larger than the House Mouse and has a very long tail. It is known to eat seeds, fruits and fibrous green plant material. It probably spends the day in a burrow.

Fragmentary evidence suggests that litters of up to four young are born around October.

HABITAT: cool-temperate semiarid woodland and shrubland on sandy clay or loam
HEAD AND BODY: 9–11 cm
TAIL: 12–14 cm
DISTRIBUTION: 10,000–30,000 km²
ABUNDANCE: very sparse
STATUS: vulnerable

(H & J Beste)

Hastings River Mouse

Pseudomys oralis (o-rah'-lis: "notable-mouth false-mouse", significance unknown)

This is a large member of the genus, about five times the weight of a House Mouse and larger than the smallest of the Australian species of *Rattus*. Its diet in the wild is not known but it has been kept in captivity on a diet consisting mostly of seeds.

HABITAT: well-watered dry sclerophyll forest with dense understorey of bracken
HEAD AND BODY: 13–17 cm
TAIL: 11–15 cm
DISTRIBUTION: less than 10,000 km²
ABUNDANCE: very sparse
STATUS: possibly endangered

(R & A Williams)

Pilliga Mouse

Pseudomys pilligaensis (pil'-ig-ah-en'-sis: "Pilliga false-mouse", from Pilliga Scrub, NSW)

Little is known of this species, which was described in 1980. It is only a little more than half the weight of the House Mouse and is the second-smallest species of *Pseudomys*. Its diet is unknown. It nests in a tunnel, probably communally.

Breeding takes place at least from October to February.

HABITAT: warm-temperate cypress-pine forest with heath understorey, on sand; localised
HEAD AND BODY: 6–8 cm
TAIL: 6–8 cm
DISTRIBUTION: less than 10,000 km²
ABUNDANCE: rare
STATUS: vulnerable

(R Whitford)

Shark Bay Mouse

Pseudomys praeconis (pree-koh'-nis: "herald false-mouse", after HMS Herald which surveyed Shark Bay in the nineteenth century)

Little is known of the biology of this species, which is about twice the size of the House Mouse. It was discovered on the mainland of Shark Bay but is now restricted to Bernier Island, where it is known to eat flowers and green plant material and to make tunnels and runways in spinifex tussocks and piles of seagrass on the beach.

The female rears up to four young in a litter. These cling tightly to the mother's teats and are dragged with her as she moves about. Young become independent at four to five weeks.

(B & B Wells)

HABITAT: subtropical semiarid coastal dunes
HEAD AND BODY: 8–12 cm
TAIL: 11–13 cm
DISTRIBUTION: less than 10,000 km²
ABUNDANCE: very sparse
STATUS: endangered

Heath Mouse

Pseudomys shortridgei (short'-rid-jee: "Shortridge's false-mouse", after G. C. Shortridge, British naturalist)

(JH Seebeck)

11 months. Breeding occurs from about October to January. Females usually rear about three young in a litter. One or two litters may be produced in a year.

HABITAT: cool-temperate heathland, particularly when regenerating after fire
HEAD AND BODY: 9–12 cm
TAIL: 8–11 cm
DISTRIBUTION: 30,000–100,000 km²
ABUNDANCE: sparse
STATUS: possibly endangered

Although almost rat-sized (and referred to by some as the Heath Rat), the Heath Mouse is a typical member of the genus *Pseudomys*. In the warmer part of the year it eats seeds, berries and flowers; in autumn it feeds on the less nutritious stems and leaves of grasses, sedges and other vegetation; in winter it survives on underground fungi.

Although mainly nocturnal, it may also be active during part of the day. It sleeps in a nest which may be built above ground among vegetation or in a burrow made by another animal. It colonises suitable areas as these begin to regenerate after fire; populations decline as regeneration slows down.

Sexual maturity is reached at about

Genus Zyzomys

(ziz'-oh-mys: significance unknown, possibly nonsensical)

The three species of rock-rats are small to medium-sized rodents with a blunt ("Roman-nosed") snout and protuberant eyes. The fur has a somewhat spiny appearance and the lightly furred tail is often swollen towards the base, owing to storage of fat. As the common name suggests, rock-rats are terrestrial; the required habitat appears to be fractured rock or rock-piles which provide long, narrow crevices in which the animals take shelter and build their nests. The female has four teats; the gestation period in the Common Rock-rat is 35 days.

Common Rock-rat

Zyzomys argurus (arg-ue'-rus: "silver-tailed *zyzomys*")

As its name implies, this species inhabits rocky areas, particularly where outcrops, screes or rock-piles create cavities in which nests can be made. It feeds on seeds, leaves and other vegetation, supplemented by insects and fungi.

Sexual maturity is reached at five to six weeks. Breeding may occur at any time of the year but appears to be greatest from March to May, in the early part of the dry season. The female rears one to four young, which are left in the nest while she forages. Young become independent at four to five weeks.

HABITAT: non-arid tropical Australia where rocks provide appropriate cavities for nesting
HEAD AND BODY: 8–10 cm
TAIL: 10–13 cm
DISTRIBUTION: 300,000–1 million km²
ABUNDANCE: common
STATUS: probably secure

(*BG Thomson*)

Central Rock-rat

Zyzomys pedunculatus (ped-unk'-ue-lah'-tus: "swollen-based [tail] *zyzomys*")

This species appears to be very rare. Only six specimens have been collected between its discovery in 1896 and its last confirmed sighting in 1960. Its habitat is much more arid than that of the other rock-rats and its biology is unknown. The specific name refers to a thickening of the tail by an accumulation of fat in the region just behind its base (giving the appearance of a very slender carrot). This is characteristic of the genus.

HABITAT: tropical arid woodland and shrubland
HEAD AND BODY: 11–14 cm
TAIL: 11–14 cm
DISTRIBUTION: less than 10,000 km²
ABUNDANCE: rare
STATUS: possibly endangered

Large Rock-rat

Zyzomys woodwardi (wood'-wud-ee: "Woodward's *zyzomys*", after B. M. Woodward, first curator of the Western Australian Museum)

The Large Rock-rat (which also occurs in New Guinea) appears to be similar in its biology to the Common Rock-rat but more selective in

(AC Robinson)

its habitat. Limited information on its diet suggests that it feeds on the fallen seeds and fruits of rainforest trees and that it nests in deep crevices among rocks.

Sexual maturity is reached at five to six weeks. Breeding appears to be possible throughout the year but to be dependent upon rainfall. The peak of breeding is in March, at the end of the wet season. One to three young are reared, becoming independent at three to four weeks.

HABITAT: tropical monsoon forest, dry sclerophyll forest and pandanus

scrubland over fractured rock, scree or rock-piles
HEAD AND BODY: 11–17 cm
TAIL: 8–12 cm
DISTRIBUTION: 100,000–300,000 km^2
ABUNDANCE: very sparse
STATUS: probably secure

Tribe HYDROMYINI

(hie'-droh-mie-een'-ee: "*Hydromys*-tribe")

Members of this tribe are carnivorous. They either are aquatic or live near water. There are numerous species in New Guinea but Australia has only two species, one widespread, the other very limited.

Genus Hydromys

(hie'-droh-mis: "water-mouse")

This genus has several species in New Guinea but only one in Australia. The hind feet are large and partially webbed. The head is rather long and somewhat flattened, with small eyes and ears and many large whiskers on the snout. The tail is short and stout. There are only two molars and the crowns of these are more or less flat, with basin-shaped depressions. Water-rats are amphibious and carnivorous.

Water-rat

Hydromys chrysogaster (kris'-oh-gas'-ter: "golden-bellied water-mouse")

The Water-rat is an opportunistic predator which takes most of its food under water. It feeds on crabs, crayfishes, frogs, small tortoises, mussels, young aquatic birds and large

(H Millen)

insects. On land, it eats food scraps and carrion and may catch small rodents and birds, including domestic poultry. It is an excellent swimmer, propelling itself with its partially webbed hind feet: the dense fur (for which it was trapped in large numbers during the nineteenth century) is water-repellent. Unlike most native rodents, it is not strictly nocturnal: it often hunts after dawn and at dusk. During the greater part of the day it sleeps in a nest at the end of a burrow dug in the bank of a watercourse or lake.

Sexual maturity is reached at about one year. Breeding occurs throughout the year but with a peak from September to March. The female has four teats and usually rears three or four young which are suckled for four weeks and remain with the mother

for a further four weeks (during which period they may learn some hunting skills from the mother). Several litters may be reared in a year.

HABITAT: close to fresh water, from tropical rainforest to cool-temperate sclerophyll forest
HEAD AND BODY: 23–37 cm
TAIL: 23–33 cm
DISTRIBUTION: more than 1 million km^2
ABUNDANCE: sparse
STATUS: probably secure

Genus *Xeromys*

(ksee'-roh-mis: "dry-mouse")

The single species of this genus is restricted to Australia, where it has one species. Like *Hydromys*, it has a rather long and somewhat flattened head with small eyes and ears, but its hind feet are not webbed. There are only two molars in each jaw and these are similar in structure to those of *Hydromys*. It is amphibious and carnivorous.

False Water-rat

Xeromys myoides (mie'-oy-dayz: "mouse-like dry-mouse")

Little is known of this elusive species, which is the only close Australian relative of the Water-rat. It inhabits rainforest and mudflats in mangrove swamps, where it has been seen to feed on soldier crabs. Its diet probably includes a much wider range of prey, including some taken in the water, although it does not have webs on the hind feet. It is known to be active in the early morning and evening, even in the middle of the day. The nest is sometimes in a burrow, sometimes on the ground, sometimes on a mound of clay that raises it above high-tide level.

The female has four teats. Nothing is known of its breeding biology.

HABITAT: mangrove mudflats and rainforest edges, close to water
HEAD AND BODY: 11–13 cm
TAIL: 9–10 cm
DISTRIBUTION: 30,000–100,000 km²
ABUNDANCE: rare
STATUS: vulnerable

(*T Redhead*)

Tribe UROMYINI

(yue'-roh-mie-een'-ee: "*Uromys*-tribe")

These are the mosaic-tailed rats, so called because the scales on the very sparsely haired tail fit together like the tiles of a mosaic, rather than overlapping. In Australia, the group is represented by the genera *Melomys* and *Uromys*. Provisionally, the prehensile-tailed rats of the genus *Pogonomys* are included in this group.

Genus *Melomys*

(mel'-oh-mis: "Melanesian-mouse")

This genus is mainly Melanesian in distribution but four or five species occur in Australia. The ears are short and the fur is long and soft. There are three molars in each jaw. The gestation period of *M. cervinipes* is 38 days, very long for an Australian native rodent.

Melomyses are primarily arboreal, climbing being aided by a short, broad hind foot and prehensile tail. Nevertheless, many species spend a great deal of time on the ground.

Grassland Melomys

Melomys burtoni (ber'-tun-ee: "Burton's Melanesian-mouse", after W. Burton, assistant to T. Bowyer-Bower, discoverer of species)

The Grassland Melomys lives in areas where grass is plentiful and feeds largely on grass seeds and succulent stems. Berries are also eaten. It has become a pest of sugarcane plantations, where it nibbles the canes, weakening them so that they become infected or break. It ascends on stout grass stems but seldom climbs trees, being an essentially terrestrial species. It usually makes a globular nest of woven grass, attached to several tall grass stems.

Breeding can occur at any time of the year but appears to be mainly from about March to July. The female has four teats and normally rears two or three young which are weaned at three or four weeks. Young attach themselves firmly to the mother's teats and are dragged behind her as she moves about.

HABITAT: subtropical to tropical grassland
HEAD AND BODY: 12–14 cm
TAIL: 12–14 cm
DISTRIBUTION: 100,000–300,000 km²
ABUNDANCE: abundant
STATUS: secure

Cape York Melomys

Melomys capensis (kay-pen'-sis: "Cape Melanesian-mouse", referring to Cape York Peninsula)

This species is indistinguishable in appearance from the Fawn-footed Melomys but is readily separated on genetic and biochemical criteria. It is nocturnal, feeding at the edges of rainforest on the fallen seeds of trees.

The female has four teats. Nothing is known of its breeding biology.

HABITAT: edges of wet sclerophyll forest and rainforest
HEAD AND BODY: 11–14 cm
TAIL: 12–18 cm
DISTRIBUTION: 30,000–100,000 km²
ABUNDANCE: rare
STATUS: vulnerable

149

Fawn-footed Melomys

Melomys cervinipes (ser-vin'-i-pez: "fawn-footed Melanesian-mouse")

The Fawn-footed Melomys is a largely arboreal rodent which spends some time on the ground. It is an excellent climber in the canopy but is unable to climb smooth-barked trees: it usually moves between the ground and the canopy along vines that surround the trunks. It is herbivorous, with a diet that includes succulent leaves and fruits. It is nocturnal and sleeps during the day in a nest of leaves constructed in a forked branch in the canopy.

Breeding occurs throughout the year, with a peak from September to June. Males are larger than females. The female has four teats and usually rears two young which are weaned at the age of three weeks. As many as five litters may be reared in a year. Suckling young attach themselves firmly to the mother's teats and are dragged behind her when she moves about.

HABITAT: subtropical to tropical wet sclerophyll forest and rainforest
HEAD AND BODY: 10–20 cm
TAIL: 11–22 cm
DISTRIBUTION: 100,000–300,000 km^2
ABUNDANCE: common
STATUS: vulnerable

(AC Robinson)

Thornton Peak Melomys

Melomys hadrourus (had-rue'-rus: "strong-tailed Melanesian-mouse")

(H & J Beste)

Largest of the Australian melomyses, this species is limited to the upper levels of Thornton Peak in north-eastern Queensland. Its biology is almost unknown.

HABITAT: montane rainforest
HEAD AND BODY: 17–18 cm
TAIL: 18–20 cm
DISTRIBUTION: less than 10,000 km^2
ABUNDANCE: rare
STATUS: vulnerable

Bramble Cay Melomys

Melomys rubicola (rue-bik'-oh-lah: "bramble-dwelling Melanesian-mouse")

There is some doubt about the validity of this species, which may be no more than a variant of the Cape York Melomys. Its major interest lies in its isolation on Bramble Cay, which is rapidly disintegrating—thus providing a literal example of "loss of habitat". It is nocturnal, probably feeding on ground vegetation and insects. It sleeps by day in a nest on the ground.

HABITAT: high grass near beach of coral cay
HEAD AND BODY: *c* 15 cm
TAIL: *c* 18 cm
DISTRIBUTION: less than 10,000 km²
ABUNDANCE: sparse
STATUS: endangered

Genus *Uromys*

(yue'-roh-mis: "tailed-mouse")

Several species of this genus of giant rats occur in Melanesia but only one in Australia. The scales of the tail are arranged in a mosaic pattern as in *Melomys* but are more prominent. The ears and eyes are small, the hind foot is large and the fur is coarse. Giant rats are primarily arboreal. The Australian species, *U. caudimaculatus*, is as large as a Water-rat; the female has four teats; the gestation period is 36 days.

Giant White-tailed Rat

Uromys caudimaculatus (kaw'-dee-mak'-ue-lah'-tus: "spotted-tailed tailed-mouse")

(AC Robinson)

HABITAT: tropical wet sclerophyll forest and rainforest
HEAD AND BODY: 25–36 cm
TAIL: 24–34 cm
DISTRIBUTION: 30,000–100,000 km²
ABUNDANCE: common
STATUS: secure

This large rodent, an excellent climber, feeds largely on the seeds and fruits of rainforest trees. It also eats insects and fungi and often raids camps or homesteads in search of food (including cans, which it can open with its incisors!). In New Guinea, it is a pest of coconut plantations. It is nocturnal and usually sleeps by day in a nest in a tree-hole, but it sometimes makes a burrow.

Males are larger than females. Sexual maturity is reached at about 10 months. Mating occurs in October or November. The female has four teats and usually rears two or three young. These attach themselves firmly to the teats and are dragged behind the mother as she moves about. They are weaned at about five weeks but do not leave the mother until about eight weeks old.

Genus *Pogonomys*

(poh-gon'-oh-mis: "bearded-mouse")

This is an essentially Melanesian genus with a number of species, one of which is represented in northern Queensland by a few scattered specimens. Prehensile-tailed rats are characterised by a short, broad head and a prehensile tail about one and a half times the length of the body. The tail appears to be naked but is very lightly haired. Females have six (sometimes eight) teats. If, as the weight of evidence suggests, *Pogonomys* is a member of the subfamily Hydromyinae, it is the only one not to have four teats.

Prehensile-tailed Rat

Pogonomys mollipilosus (mol'-ee-pil-oh'-sus: "soft-furred bearded-mouse")

The identity of the prehensile-tailed rat found in Australia is not firmly determined but is provisionally assigned to *P. mollipilosus*. The common name refers only to the genus. In New Guinea, this species feeds on leaves and nuts, both on the ground and in trees, where it is an agile climber. It nests communally in burrows.

Nothing is known of the pattern of breeding in Australia. Although the female has six teats, she usually rears only two or three young.

HABITAT: tropical rainforest
HEAD AND BODY: 13–15 cm
TAIL: 16–21 cm
DISTRIBUTION: less than 10,000 km²
ABUNDANCE: rare
STATUS: possibly endangered

(H & J Beste)

Subfamily MURINAE

(myue-reen'-ee: "mouse-subfamily")

The Murinae comprises more than four hundred species of rats and mice, occurring in Eurasia, Africa, Melanesia and Australia. However, of some 90 genera in the group, only one, *Rattus*, occurs in Australia. The native species of this genus appear to have come to Australia less than 1 million years ago and some of these may be much more recent arrivals. Two species of *Rattus*, and the House Mouse, *Mus musculus*, were introduced by European explorers and settlers within the last three hundred years.

Genus *Rattus*

(rat'-us: "rat")

This large, cosmopolitan genus is represented by seven native species in Australia. They are perhaps best described as having a "rat-like" appearance, with a somewhat "Roman" profile to the head, large eyes and an almost naked tail with rings of scales. With the exception of the Long-haired Rat, which periodically invades desert regions after several years of unusually good rainfall, members of this genus are restricted to well-watered parts of Australia.

Dusky Rat

Rattus colletti (col'-et-ee: "Collett's rat", after R. Collett, Norwegian zoologist)

This is a species of tropical monsoonal flood plains. During the floods of the wet season, it moves to higher ground, feeding on the bases of grasses and the corms of sedges. As the floods subside, it spreads out over the plains, feeding on sedges and sheltering in the cracks that develop in the drying soil. It is an opportunistic species: populations increase greatly when food is available but there is usually a heavy mortality in the dry season.

Females become sexually mature when about six weeks old. They have 12 teats and usually rear about nine young in a litter. These become independent at about three to four weeks. Breeding can continue throughout the year but has a peak from May to June.

HABITAT: tropical flood plains with grassy borders
HEAD AND BODY: 7–21 cm
TAIL: 8–15 cm
DISTRIBUTION: 100,000–300,000 km²
ABUNDANCE: common
STATUS: vulnerable

(HJ Aslin)

Bush Rat

Rattus fuscipes (fus'-kee-pez: "dusky-footed")

(AC Robinson)

teats and usually rears about five young in a litter: these become independent when four to five weeks old.

HABITAT: cool-temperate to tropical coastal rainforest
HEAD AND BODY: 11–20 cm
TAIL: 10–20 cm
DISTRIBUTION: 300,000–1 million km²
ABUNDANCE: abundant
STATUS: secure

The Bush Rat is a very adaptable species which feeds upon a wide range of green plants and fungi and has a considerable intake of insects. Being dependent upon moist conditions and daily access to drinking water, it is restricted to well-watered coastal regions, usually forests with dense ground cover. It sleeps by day in short burrows or under the shelter of stones or fallen timber. Local populations reach greatest density in regenerating bushland two to three years after a severe fire.

Males are larger than females. Females become sexually mature when four to five months old and are thereafter able to breed continuously, with a peak of births in summer. The female has eight

Cape York Rat

Rattus leucopus (lue'-koh-poos: "white-footed rat")

Like the Dusky Rat, this is a tropical species, very dependent upon a wet environment. It feeds at night on the fruits and nuts of rainforest trees, supplemented by insects; it does not climb trees. During the day it sleeps in a nest under a log or among buttress roots of rainforest trees.

Sexual maturity is reached at about three months of age. Females have six teats and usually rear three or four young, which become independent at about four weeks. Breeding extends throughout the year except for the dry season.

HABITAT: tropical rainforest
HEAD AND BODY: 15–21 cm
TAIL: 14–21 cm
DISTRIBUTION: 30,000–100,000 km²
ABUNDANCE: abundant
STATUS: probably secure

(AC Robinson)

Swamp Rat

Rattus lutreolus (lue'-tree-oh'-lus: "otter-like rat")

Both the scientific and common names of this species were based on the misapprehension that it is amphibious. In fact, its preferred habitat is dense ground vegetation. It constructs runways through the vegetation and usually makes its nest at the end of a burrow. In swampy areas, nests are built above ground, in hollow logs or in tussocks. The

major components of the diet are grasses and sedges, including their seeds. Mosses and fungi are also eaten and insects are taken, particularly in summer. Population densities are higher in areas where vegetation is regenerating several years after a bushfire.

Females become sexually mature at about three months. Breeding

occurs throughout the year except in winter; the season is longer in the northern part of the range than in the southern. Females from the mainland have 10 teats; those from Tasmania have eight. The average litter is three or four young which are weaned at the age of about three weeks.

HABITAT: tropical to cool-temperate dense grassland, sedgeland or heath
HEAD AND BODY: 13–20 cm
TAIL: 8–15 cm
DISTRIBUTION: 100,000–300,000 km²
ABUNDANCE: abundant
STATUS: secure

(R Miller)

Canefield Rat

Rattus sordidus (sor'-did-us: "dirty rat")

Although now common in sugar canefields, this species was originally an inhabitant of tropical grasslands and grassy areas within and at the edges of forests. However, sugarcane (which is a grass) provides excellent food and shelter. The normal diet is grass, but this is supplemented to some extent by insects. During the day, it sleeps in a nest, usually situated in a burrow.

Males are larger than females. Sexual maturity is reached at nine to 10 weeks. Breeding can occur throughout the year but there is a peak of births from about March to May. The female has 12 teats and usually rears about six young in a litter. Since these become independent when about three weeks old, populations can expand very rapidly.

HABITAT: tropical grasslands and canefields
HEAD AND BODY: 11–21 cm
TAIL: 10–20 cm
DISTRIBUTION: 100,000–300,000 km²
ABUNDANCE: common
STATUS: secure

(R & A Williams)

Pale Field-rat

Rattus tunneyi (tun'-ee-ee: "Tunney's rat", after J. T. Tunney, collector of first specimen)

The Pale Field-rat feeds mainly at night on the seeds, stems and roots of grasses. During the day it sleeps in a nest in a shallow burrow, possibly communally.

Sexual maturity is reached at about five weeks. Breeding extends from about May to August in the northern part of the range, possibly from March to May in the southern part. The female has 10 teats and usually rears a litter of four, which become independent at about three weeks.

HABITAT: subtropical to tropical tall grassland, usually close to water
HEAD AND BODY: 12–20 cm
TAIL: 8–15 cm
DISTRIBUTION: 100,000–300,000 km²
ABUNDANCE: common
STATUS: secure

(B & B Wells)

155

Long-haired Rat

Rattus villosissimus (vil'-o-sis'-im-us: "extremely hairy rat")

Unlike other native rats, this species ventures into desert regions, but only following one or two years of unusually heavy rains that lead to abundant plant growth. Under these circumstances it multiplies rapidly to "plague" proportions, after which populations crash and it survives, in much lower density, in better-watered areas at the edge of deserts. It is not desert-adapted, since it requires regular access to drinking water. It feeds, usually at night, on grasses, seeds, succulent plants and insects; during plagues, it will attempt to eat any organic material, boldy entering houses or camps to do so. Under normal circumstances it sleeps in a nest in a burrow forming part of a warren.,

Males are a little larger than females. Sexual maturity is reached at about 10 weeks. Breeding can occur at any time of the year but appears to be related to the availability of food. The female has 12 teats and usually rears about seven young in a litter.

HABITAT: semiarid to arid areas over most of inland Australia, on an opportunistic basis
HEAD AND BODY: 12–22 cm
TAIL: 10–18 cm
DISTRIBUTION: more than 1 million km^2
ABUNDANCE: rare to abundant
STATUS: secure

(H J Aslin)

Order CARNIVORA

(kar-niv'-or-ah: "flesh-eaters")

This large group of mammals includes cats, dogs, bears, weasels, otters and two groups of marine mammals, commonly known as "seals". It was once thought that "seals" comprised a natural grouping (i.e. that they had a common seal-like ancestor) but it is now apparent that there are two groups which, quite separately, became adapted to life in the sea.

Members of one group, the superfamily Otaroidea, retain small external ears and have hind limbs that can be moved forward or backward of the hip to contribute to a shuffling walk. This group includes two families: the Otariidae (sea-lions and fur-seals) and the Odobenidae (walruses).

In the other group, the Phocoidea, there is no external ear flap and the hind limbs are permanently directed backwards to create the equivalent of a tail fin. The group has only one family, the Phocidae ("true" seals), none of which are resident in Australian coastal waters.

Family CANIDAE

(kah'-nid-ee: "*Canis*-family")

These carnivores form a relatively small family of rather similar animals, including foxes, jackals, wolves and dogs. The only canid living in Australia before European settlement was the Dingo. Subsequently, the Red Fox was introduced.

Genus Canis

(kah'-nis: "dog")

This genus includes the wolves, the coyote, jackals and dogs. Species of the genus occur on all continents.

Dingo

Canis familiaris (fah-mil'-ee-ah'-ris: "domestic dog")

The Dingo belongs to the same species as the domestic dog and interbreeds freely with it. Nevertheless, it differs sufficiently to be regarded as a subspecies, *Canis familiaris dingo*. It does not bark and it breeds only once a year. The closest relatives of the Dingo are the wild or semi-domesticated dogs of South-East Asia, Indonesia and New Guinea. It is often said that Aborigines brought the Dingo to Australia but, since it appears to have been on the continent for no more than 6000 years, this seems very unlikely.

It is an opportunistic predator on mammals ranging in size from rodents to kangaroos; sheep and calves are also taken. When small prey are plentiful, Dingos are solitary hunters but, when these are scarce and larger prey must be tackled, cooperative packs may form.

Although its overall range includes the whole of the continent (but not Tasmania), it is not well adapted to arid conditions and usually needs to drink once a day.

Sexual maturity is reached at about one year. Mating occurs from about March to June and a litter of three or four young is born from about August to November.

HABITAT: almost any part of the Australian mainland which provides access to drinking water, but preferably woodland and grassland adjacent to forest
HEAD AND BODY: 85–100 cm
TAIL: 25–38 cm
DISTRIBUTION: more than 1 million km²
ABUNDANCE: common
STATUS: secure

(R Strahan)

Family OTARIIDAE

(oh'-tah-ree'-id-ee: "*Otaria*-family" after a genus of eared seals)

The eared-seals are usually divided into two subgroups: so-called sea-lions have a sleek coat; fur-seals have a much thicker underfur and a slightly more shaggy appearance. The distinction is artificial.

All otariids propel themselves through the water by strokes of their forelimbs ("flippers").

Genus Arctocephalus

(ark'-toh-sef'-al-us: "bear-head")

This is the largest genus of the fur-seals. It is restricted to the Southern Hemisphere but one species occurs in the Galàpagos. Members of this genus have thick fur.

New Zealand Fur-seal

Arctocephalus forsteri (for'-ster-ee: "Forster's bear-head", after George Forster, assistant naturalist and artist on Cook's voyage in HMS Resolution)

Widely distributed in New Zealand, this species also occurs in the Great Australian Bight. It feeds on squids and barracuda which live near the surface but also dives deep for octopuses and spiny lobsters. It is known also to take penguins.

Males are much larger than females. Females give birth to a single young between November and January. Pups are suckled for 10 to 11 months. Mating occurs about one week after a birth.

HABITAT: cool coastal waters with undisturbed rocky beaches: breeding animals require access to vegetation for shelter, and rock pools for cooling
OVERALL LENGTH: 1.5–2.5 m (males), 1.3–1.5 m (females)
DISTRIBUTION: not applicable
ABUNDANCE: very sparse
STATUS: probably secure

(GW Johnstone)

Australian Fur-seal

Arctocephalus pusillus (pue-sil'-us: "weak bear-head")

This species is largely confined to Bass Strait. It is an excellent swimmer and descends to considerable depths in search of small squids, fishes and rock lobsters. It seems probable that it finds some prey by echolocation.

Males are much larger than

females. Breeding animals come ashore on rocky coasts and births occur in November and December. Mating occurs about a week after a female has given birth to her single young, which is not weaned until 10 to 11 months old.

HABITAT: cool-temperate continental shelves with rocky shores
OVERALL LENGTH: 2.0–2.3 m (males), 1.3–1.7 m (females)
DISTRIBUTION: not applicable
ABUNDANCE: common
STATUS: probably secure

(RM Warneke)

Genus Neophoca
(nee'-oh-foh'-kah: "new-seal")

This genus has only one species, restricted to Australian waters. Its underfur is less dense than in fur-seals of the genus *Arctocephalus*.

Australian Sea-lion
Neophoca cinerea (sin'-e-ray'-ah: "ash-coloured new-seal")

This species lives mostly on the Western Australian and South Australian coasts and forms breeding colonies on rocky shores. Males are distinguished by a cape of pale hair over the back of the head and shoulders. The diet appears to consist mostly of squids and cuttlefishes but fishes and rock lobsters are also eaten.

The male is much larger than the female. Births seem usually to occur from October to January, but there is some evidence from Kangaroo Island of an 18-month breeding cycle, with a summer peak of births, a year with no births and then births in midwinter. If this is a normal phenomenon, it defies explanation. Mating takes place about one week after a female has given birth to a single young.

HABITAT: subtropical to cool-temperate continental shelves with rocky shores for breeding
OVERALL LENGTH: 1.9–2.4 m (males), 1.6–1.7 m (females)
DISTRIBUTION: not applicable
ABUNDANCE: very sparse
STATUS: probably secure

(A Eames)

Order SIRENIA

(sie-ray'-nee-ah: "sirens", in the sense of mythical alluring semihuman females)

These are the only herbivorous aquatic mammals: they graze upon sea-grasses and other aquatic vegetation in shallow tropical to subtropical coastal waters, estuaries or rivers. They have no hind limbs and the tail supports a whale-like horizontal fluke. The forelimbs are in the form of rounded flippers. Females have two teats on breast-like mammae.

The order is divided into two families, the Trichechidae which includes the manatees, and the Dugongidae with a single species, the Dugong.

Family DUGONGIDAE

(dyue-gong'-id-ee: "*Dugong*-family")

Members of this group resemble manatees (family Trichechidae) in general appearance but differ in many respects. The tail-fluke of the Dugong has a concave trailing edge (round in manatees); males have short tusk-like incisor teeth (absent in manatees); and the molars are reduced to a few peg-like rudiments (well-developed series which are continually replaced in manatees). Dugong are distributed through the Indian Ocean and western Pacific; manatees are restricted to the Atlantic Ocean.

Genus Dugong

(dyue'-gong: "*dugong*", from Malay, duyong, name for this animal)

The characteristics of the genus are those of the species.

Dugong

Dugong dugon (dyue'-gon: "*dugong dugong*")

The Dugong is an aquatic herbivore which feeds on sea-grass in shallow tropical seas. It swims relatively slowly by means of vertical movements of its horizontally fluked tail, coming to the surface to breathe through two valvular nostrils at the front of the head. The forelimbs are in the form of flippers and hind limbs are absent.

There is no definite breeding season and up to six years may elapse between births. The single young accompanies its mother for up to two years.

HABITAT: shallow tropical seas with sandy bottom and growth of sea-grasses
OVERALL LENGTH: up to 3 m
DISTRIBUTION: not applicable
ABUNDANCE: very sparse
STATUS: vulnerable

(*B Cropp*)

Introduced Species

Order PERISSODACTYLA

(pe'-ris-oh-dak'-til-ah: "odd-toed")

The horses, rhinoceroses and tapirs that comprise this group have feet in which the middle (third) toe takes most or all of the weight of the body.

Family EQUIDAE

(ek'-wid-ee: "*Equus*-family", after the genus of horses and donkeys)

The feet of horses, donkeys and zebras are reduced to a single (third) toe, which bears a large hoof. The seven species in the genus are grazing animals which crop grasses with their upper and lower incisor teeth.

Genus Equus

(ek'-wus: "horse")

The characteristics of the genus are those of the family.

Donkey

Equus asinus (ah-seen'-us: "ass horse")

Domesticated for at least 6000 years, the Donkey appears to be native to north-eastern Africa. It was introduced to Australia as a pack and draught animal, better adapted to arid conditions and rough terrain than the Horse. However, when motor transport became readily available, its use declined rapidly. Released to fend for themselves, Donkeys became feral over most of the drier inland region and the rugged monsoon country of the Top End and Kimberley. Despite intensive control measures, the population appears to be between one and two million. It associates in herds of up to 30 when food is plentiful but may form groups of several hundred when watering points are limited during the dry season or in times of drought. It is a grazer which also browses on a wide range of shrubs. It is able to drink brackish water.

Females become sexually mature when about two years old. Males are usually older before they have the opportunity to mate. Breeding can occur at any time of the year. One young is born after gestation of 53 to 54 weeks.

HABITAT: temperate to tropical woodland, shrubland, tussock and hummock grassland, gibber desert, rocky dissected country

HEAD AND BODY LENGTH: to 2.5 m

TAIL LENGTH: to 30 cm

DISTRIBUTION: more than 1 million km^2

ABUNDANCE: common

STATUS: secure

(*R. Strahan*)

Horse

Equus caballus (kah-bal'-us: "horse horse")

Populations of the feral Horse, or Brumby, are derived from stock that have escaped or have been deliberately let loose to breed since the late 18th century. The Horse is primarily a grazer but may browse to a limited extent. Access to drinking water is required at intervals of no more than four days. It is social, usually moving in groups comprising a dominant male, up to six mature females, and their immature offspring. Temporary aggregations of up to 100 may form where food is abundant or water is scarce.

Females become sexually mature at 18 to 24 months; males seldom have the opportunity to mate until at least five years old. Courtship is cursory. After a gestation of 47 to 48 weeks, a female separates from the social group and gives birth to a single young in seclusion, returning with it after one to two weeks. When a female reaches maturity she may remain in the parental group or join another. Mature males are ejected from the parental group and lead a solitary existence until they are able to make a successful challenge for supremacy of an established group.

HABITAT: mainly grassland but also open woodland (including sub-alpine areas), with access to drinking water
HEAD AND BODY LENGTH: to 2 m
TAIL LENGTH: to 40 cm
DISTRIBUTION: more than 1 million km²
ABUNDANCE: common
STATUS: secure

(JE Wapstra)

Order ARTIODACTYLA

(art'-ee-oh-dak'-til-ah: "even-numbered toes")

The pigs, camels, deer, giraffes, antelopes, sheep and cattle that comprise this large group are all hoofed mammals which (in contrast to the Perissodactyla) have an even number of toes on each foot: the weight of the body is taken mostly on the equally developed third and fourth toes. Most artiodactyls are grazers and/or browsers but pigs are omnivorous. Families of the Artiodactyla represented by feral species in Australia are the Suidae (Pig), Camelidae (Camel), Cervidae (deer) and Bovidae (cattle and Goat).

Family BOVIDAE

(boh'-vid-ee: "*Bos*-family", after a genus of cattle)

Members of this large group of artiodactyls include the sheep, goats and cattle as well as the Pronghorn Antelope, two species of duikers and 11 genera of spiral-horned antelopes. They are characterised by having two hooves on each foot and a multichambered stomach that efficiently digests fibrous plant material. The muzzle is moist. Males (and the females of some species) have horns: in contrast to the annual, bony antlers of deer, the horns are permanent and composed of keratin, the same substance as claws, nails and hooves. Of the 22 genera distributed over Eurasia, Africa and North America, only three are represented in Australia.

Given that cattle may be allowed to range freely between musters and that musters are not always complete, it is sometimes difficult to distinguish between feral and managed populations. The view is taken here that only the Water Buffalo and (some) Banteng are feral in Australia. Many Goat populations are clearly feral but it is doubtful that any Sheep exist in a truly feral state.

Genus Bos
(bos: "ox")

Members of this genus have a blunt, moist snout. The horns, set just behind the ears, are round to oval in cross-section.

Banteng
Bos javanicus (jah-vahn'-ik-us: "Javan ox")

The Banteng, which interbreeds freely with domestic cattle (*Bos taurus*), is native to Indonesia. It was introduced from Bali to the Top End in 1849 and the feral population remains restricted to the Cobourg Peninsula. Very few pure stock remain outside captivity. It is rather delicately built and attractively marked, with pale "boots" and has a pale patch on the hindquarters. During the day it rests in monsoon forest, emerging at night to graze—and to some extent browse—in more open country. It is able to drink brackish water.

The Banteng usually moves about in mixed herds of up to 30 cows, calves and juvenile males. Older males tend to remain outside the herds, visiting them during the mating season. A single young is born after a gestation of 38–40 weeks.

HABITAT: tropical woodland
HEAD AND BODY LENGTH: 1.8–2.0 cm
TAIL LENGTH: *c* 60 cm
DISTRIBUTION: 10,000–30,000 km^2
ABUNDANCE: very sparse
STATUS: vulnerable

(R Strahan)

Genus *Bubalus*

(bue-bah'-lus; "buffalo")

The two species in this genus have horizontally oriented, swept-back horns that are crescentic in cross-section.

Water Buffalo

Bubalus bubalis (bue-bah'-lis: "buffalo buffalo")

Almost extinct in the wild, this species has been domesticated for several thousand years in Asia and Indonesia. The ancestors of the Australian stock, now feral in the Top End, were domesticated animals, introduced into Australia from Timor and Indonesia in the first half of the 19th century. They are now fiercely independent, but young animals brought into captivity can readily be tamed. It is an animal of the swampy tropics and normally wallows daily in a shallow, muddy pool: its hooves are large and splayed in adaptation to its habitat. At night, it sleeps under tree cover, grazing and browsing by day. Through most of the year, animals are segregated into single-sex herds; groups of females are led by a dominant female, while males form looser groups.

During the extended mating season, beginning in summer, older males visit the female herds and attempt to mount females that are on heat. A single young is born after a gestation of 44 to 47 weeks. An interesting aspect of maternal behaviour is that young animals are formed into creches under the care of a "roster" of a few females who guard them while the other mothers go off to feed and to wallow.

HABITAT: tropical swampy woodlands and forest edges
HEAD AND BODY LENGTH: 2.5–3.0 m
TAIL LENGTH: *c* 70 cm
DISTRIBUTION: 300,000–1 million km²
ABUNDANCE: abundant
STATUS: secure

(*R Strahan*)

Genus *Capra*

(kap'-rah: "goat")

The four species in this genus are all native to mountainous regions of Eurasia and North Africa. They have short, powerful legs, the hooves of which can be employed as pincers to grip on rocky substrates. Males have large, swept-back horns; those of the females are smaller. Males have a beard.

Feral Goat

Capra hircus (her'-kus: "goat goat")

This species, known to some authorities as *C. aegagrus*, may have had its home from Pakistan to the Middle East but, having been domesticated for not less than 8000 years, its origins have been obscured. It was introduced to Australia very early in the late 18th century and rapidly established feral populations, now extending over about half of the mainland but absent from rainforests, wetlands and the driest deserts. It is primarily a diurnal browser on shrubs and low trees but also eats grasses: under pressure, a group will strip an area bare of almost all vegetation. The preferred habitat is rocky hillsides, into which

Goats escape when disturbed, but they may inhabit almost any type of land and they are able to climb into trees with low branches. There is a requirement for shelter, preferably among rocks, but also in dense vegetation.

Feral Goat populations usually move about in groups of up to 20 but larger aggregations may occur where food is plentiful or water is scarce. The social organisation is not well understood but it seems that, for at least part of the year, mature males are excluded from the group.

Females become sexually mature in the first year, sometimes as early as six months. Mating, which takes

place shortly after parturition, can occur at any time of the year but with a slight peak in summer and early autumn. One or two young are born after a gestation of about 21 weeks.

HABITAT: preferably temperate to subtropical rocky hillsides with vegetation ranging from shrubs to open woodland but also on plains. Absent from very arid and very wet regions.
HEAD AND BODY LENGTH: 1.1–1.6 m
TAIL LENGTH: *c* 12–17 cm
DISTRIBUTION: more than 1 million km²
ABUNDANCE: sparse to abundant
STATUS: secure

(B & B Wells)

Family CAMELIDAE

(kah-mel'-id-ee: "*Camelus*-family", after the genus of camels)

Camels and llamas are typical artiodactyls in that each foot is supported by two toes but, instead of being directed downward, these are aligned horizontally, sitting above fleshy pads which take the weight of the body: the hooves lie at the front of the pads. Camelids are ruminants, with a four-chambered stomach in which vegetable fibre is digested. They crop vegetation between their upper and lower incisors.

Genus Camelus

(kah-mel'-us: "camel")

Expert opinion differs on whether this genus includes one or two species. The Dromedary, *C. dromedarius*, has one hump; the Bactrian Camel (either *C. d. bactrianus* or *C. bactrianus*) has two. No Dromedaries exist in a natural wild state but there are some wild populations of the Bactrian Camel.

Camel

Camelus dromedarius (drom'-ed-ar'-ee-us: "dromedary camel")

The Camel was introduced into Australia in the mid-19th century, mostly from the region that is now Pakistan. Originally of great value as a draught animal in arid and semiarid Australia, Camels were rapidly replaced by motor transport in the 1920s and most were set free. Their descendants constitute the only non-domesticated population in the world.

The hump of the Camel is composed largely of fatty tissue which serves as a store of food energy when an animal is starving. Oxidation of food leads to the production of water but, since this is the case with all animals, it is not strictly relevant to say that the Camel survives under dry conditions by access to this "water of metabolism". Its success is due rather to an ability to reduce its loss of water by evaporation and to produce a highly concentrated urine and thus minimise loss of water through excretion. It can even survive under conditions when loss of body water has begun to make its blood rather viscous. When given the opportunity to drink, a dehydrated individual may drink up to 200 litres (about 20 per cent of its body weight).

The Camel lives in herds of fewer than 10 to more than several hundred. The social structure is loose, herds being led for most of the time by a senior female. Sexual maturity is reached at about four years. Breeding can occur at any time of the year but is largely determined by the abundance of food. Females

(B & B Wells)

produce a single young after a gestation of 51 to 54 weeks.

HABITAT: temperate to subtropical arid sandy deserts with succulent vegetation and shrubs
HEAD AND BODY LENGTH: *c* 3 m
TAIL LENGTH: *c* 70 cm
DISTRIBUTION: more than 1 million km^2
ABUNDANCE: sparse to common
STATUS: secure

Family CERVIDAE

(ser'-vid-ee: "*Cervus*-family", after a genus of deer)

Deer are two-hooved ruminants. Males have branched bony antlers that are grown and dropped each year. There are no upper incisors: the lower incisors (and incisor-like lower canines) bite against a tough pad at the front of the palate. Deer are native to America, Eurasia and Indonesia. Of the nine genera, three are represented as feral populations in Australia: all belong to the subfamily Cervinae.

Genus Axis

(ax'-is: possibly "deer")

The four species in the genus have a spotted coat and lyre-shaped antlers with widely separated points. Upper canines are usually lacking. The two species of *Axis* occurring in Australia are smaller than the other Australian deer.

Chital

Axis axis (ax'-is: "deer deer")

The Chital (or Spotted Deer) is notable for its dappled pattern of white spots on a light brown back and sides. A native of India, it was introduced to Australia early in the 19th century for hunting. Some large populations were developed in captivity but the species failed to establish itself except in a small area of northern Queensland. It is predominantly a grazer but also browses on shrubs.

It moves in herds of up to 100 individuals. In the mating season (February to May, August, September) males segregate from the herd but visit groups of females, attempting to copulate with those that are on heat. After a gestation of 30 to 32 weeks, a single young is born.

HABITAT: subtropical forest edges and woodland
HEAD AND BODY LENGTH: 1.3–1.9 m
TAIL LENGTH: *c* 25 cm
DISTRIBUTION: less than 10,000 km²
ABUNDANCE: very sparse
STATUS: possibly endangered

(CA Henley)

Hog Deer

Axis porcinus (por-seen'-us: "pig-like deer")

This species bears no resemblance to a pig except in its preference for swampy areas and its rather short legs: it is only 50 to 70 cm high at the shoulder (the Red Deer is up to 120 cm). Males have small and rather simple antlers. Native to India and Sri Lanka, it was introduced to Victoria in the 1860s and has become established in coastal scrubland and swamps in southern Gippsland. Primarily a grazer, but also browsing on native shrubs, it is nocturnal, moving about singly or in groups of four or less, seldom far from water, into which it may flee when disturbed.

Most mating appears to take place in February and March. Males are very aggressive at this time, attacking almost any moving object and even stationary objects such as trees. The gestation period is 28 to 32 weeks and twins are not uncommon.

HABITAT: tropical to cool temperate swampy woodland
HEAD AND BODY LENGTH: 1.2–1.4 m
TAIL LENGTH: 12–14 cm
DISTRIBUTION: 10,000–30,000 km²
ABUNDANCE: sparse
STATUS: vulnerable

(R Mayze)

Genus Cervus
(serv'-us: "deer deer")

This large genus includes the "true" deer, characterised by their medium to large size, large and heavy antlers and, often, a mane around the neck. Upper canine teeth are present.

Red Deer
Cervus elaphus (el-ah'-fus: "deer deer")

Native to Europe, western and central Asia, and north-western Africa, the Red Deer was introduced to Australia in the 1860s from semi-domesticated British stock, for the purpose of hunting. It has not adapted well but is somewhat precariously established in relatively small areas in Queensland and Victoria. It is a grazer and browser, feeding mainly in the evening or at night. Mature males have many-branched antlers which may be up to 90 cm long.

From about March to August, populations segregate into all-male groups and groups of females and their young. Between about September and December, the male herds break up. Mating occurs in March or April, accompanied by fierce combat between mature males competing for females that are on heat, each male attempting to dominate as many females as possible. After a gestation of 33 to 34 weeks, a single young is born in November or December. Young animals have a white-spotted coat.

HABITAT: temperate forest edges or woodland with grassy understorey
HEAD AND BODY LENGTH: 1.9–2.0 cm
TAIL LENGTH: 12–15 cm
DISTRIBUTION: 10,000–30,000 km²
ABUNDANCE: very sparse
STATUS: vulnerable

(D Panther)

Rusa Deer

Cervus timoriensis (tee'-mor-ee-en'-sis: "Timor deer")

This species, which occurs naturally through most of Indonesia, has been introduced to islands in Torres Strait and the Gulf of Carpentaria. A population has existed in Royal National Park, Sydney, since early in the 20th century. It is primarily a grazer. Except in the mating season, males associate in "bachelor" herds, females and juveniles remaining in separate herds.

Mating can occur at any time of the year but is mostly in winter. Mating is preceded by intense aggression between males, which carry piles of vegetation on their antlers.

HABITAT: tropical grasslands
HEAD AND BODY LENGTH: 1.4–1.9 m
TAIL LENGTH: *c* 20 cm
DISTRIBUTION: 10,000–30,000 km²
ABUNDANCE: rare
STATUS: possibly endangered

(GB Baker)

Sambar

Cervus unicolor (ue'-nee-kol'-or: "single-coloured deer")

(GB Baker)

Native to India and Sri Lanka, the Sambar was introduced into Victoria in the 1860s for hunting. It is now established over a large part of Gippsland, particularly in wet, mountainous forests. It is notable for its very large "bat-like" ears and is very shy, seldom venturing far from cover. It is a grazer and browser, feeding on a wide range of coarse grasses and native foliage. Much less social than other Australian deer, it usually associates in pairs or in family groups of three or four. It often wallows in muddy pools.

Mating occurs in September and October or in March and April. Mature males establish territories. Females aggregate into single-sex groups and, as they pass through the territory of a male, he attempts to mount any females that are on heat. A single young is born after a gestation of about 34 weeks.

HABITAT: cool temperate forest, adjacent to grassland, usually close to water
HEAD AND BODY LENGTH: 1.6–2.5 m
TAIL LENGTH: 20–30 cm
DISTRIBUTION: 100,000–300,000 km²
ABUNDANCE: very sparse
STATUS: vulnerable

Genus Dama

(dah'-mah: "deer")

The single species in this genus has a rather slender build. The many-branched antlers are flattened and webbed. It lacks upper canines.

Fallow Deer

Dama dama (dah'-mah: "deer deer")

Possibly native to the region around Turkey, the Fallow Deer was long ago translocated over much of Europe to provide hunting for aristocrats: it no longer exists in a truly wild state. The Australian Fallow Deer are derived from stock imported from Great Britain between 1830 and 1880, and isolated feral populations extend from northern Queensland to South Australia. Coloration varies but the commonest form has white spots on the pale brown back and upper sides.

Except where subject to harassment, it feeds by day, mostly as a grazer but also browsing upon shrubs and the foliage and bark of trees. During the period when males are regrowing their antlers, they associate with females in mixed herds. As the antlers become large, males segregate into bachelor groups. In the mating season, April and May, males become markedly aggressive and the successful combatants establish territories within which they dominate groups of females. After a gestation of 33 to 34 weeks, a single young is born in November or December.

HABITAT: temperate woodland and woodland edges
HEAD AND BODY LENGTH: 1.2–1.7 m
TAIL LENGTH: 19–23 cm
DISTRIBUTION: 30,000–100,000 km²
ABUNDANCE: very sparse
STATUS: probably secure

(GB Baker)

Family SUIDAE

(sue'-id-ee: "*Sus*-family," after the genus of pigs)

This small family, with only nine species in five genera, occurs naturally over most the world. The feet have four hoofed toes but the central two are more strongly developed than the outer ones, which seldom touch the ground. The canines are strongly developed in males and may form curved, upwardly directed tusks. In contrast to other artiodactyls, suids are omnivorous, preying upon a wide range of smaller vertebrates as well as eating plant material. The family is distributed over Eurasia, Indonesia and North Africa but is absent from the Americas, where the related family Tayassuidae (peccaries) is found.

Genus Sus

(soos: "pig")

The five species in this genus are distrubuted over Eurasia, Indonesia and North Africa. They are typical members of the family.

Pig

Sus scrofa (skroh'-fah: "sow-pig")

This species still exists in Eurasia and North Africa as the Wild Boar, various populations of which have been domesticated for at least 6000 years. The feral mainland Australian population is largely descended from domestic breeds brought into the country by European settlers, but it is possible that closely related species from Indonesia or New Guinea have been introduced into northern Australia and have hybridised with the European stock.

It is an animal of well-watered (usually flood-prone) country with at least some dense vegetation. It feeds on succulent roots, grain crops, berries and fruits, frogs, lizards, small to medium-sized vertebrates and carrion. It rests by day in an ill-defined dust wallow, usually in the shelter of vegetation. It is nocturnal, with most activity around dawn and dusk. It usually begins the night's activity by wallowing in a muddy pool and much of its environmental damage is caused by this habit.

Pigs are solitary and territorial. Females become sexually mature at about eight months and usually produce a litter of about six young. Gestation is about 16 weeks and a female often rears two litters in a year. Males play no parental role.

HABITAT: Almost any temperate or tropical environment with dense vegetation and shallow water
HEAD AND BODY LENGTH: 1.0–1.5 m
TAIL LENGTH: 25–30 cm
DISTRIBUTION: more than 1 million km^2
ABUNDANCE: common
STATUS: secure

(B & B Wells)

Order RODENTIA

See page 126.

Order CARNIVORA

See page 156.

Family FELIDAE

(fel'-id-ee: "*Felis*-family", after a genus of cats)

About 20 genera, ranging in size from the domestic Cat to the Lion, comprise this family, sharing, among other characteristics, short, broad heads and sharp, retractile claws. Although primarily terrestrial, many are excellent climbers.

Genus **Felis**

(fel'-is: "cat")

The six species in this genus are similar, in all but minor respects, to the domestic Cat.

Cat

Felis catus (kah'-tus: "cat cat")

(IT Mahood)

of two to seven (usually four to five) after a gestation of six weeks. Young may remain with the mother until three to seven months old.

HABITAT: tropical rainforest to desert and alpine heath; towns and cities
HEAD AND BODY LENGTH: 38–63 cm
TAIL LENGTH: 23–34 cm
DISTRIBUTION: more than 1 million km^2
ABUNDANCE: abundant
STATUS: secure

Brought to Australia as domesticated animals, this species soon spread all over Australia. It preys on mammals varying in size from the Rabbit to the smallest rodents and marsupials. Birds are seldom an important part of the diet except on islands and in rookeries: in arid regions, lizards feature largely in the diet. The Cat does not need to drink, obtaining sufficient water from its prey. It is nocturnal, spending the day in a den or in dense undergrowth.

It is solitary and territorial. Females become sexually mature at the age of 10 to 12 months, males several months later. Mating may occur at any time of the year, leading to a litter

Genus *Vulpes*

(vool'-payz: "fox")

This genus includes eight species of foxes, distributed through Eurasia, Asia and North Africa. It is closely related to *Canis*, differing mainly in behaviour and minor details of the anatomy of the skull.

Red Fox

Vulpes vulpes (vool'-payz: "fox fox")

The Red Fox was introduced into Australia in the 19th century to be hunted on horseback. It has spread over most of the southern two-thirds of Australia, but not Tasmania. It preys on the Rabbit, rodents and birds but also eats carrion, large insects, berries and fruits. It usually spends the day in a den but may be active by day in cold weather.

Sexual maturity is reached in the first year. Mating occurs in June or July and the first mating usually leads to a monogamous union. The male brings food to his mate, and the parents cooperate in rearing the weaned cubs. They are often assisted by elder offspring that are mature but sexually inactive for a year or more. About four young are born in late winter or early spring.

HABITAT: temperate to subtropical forest to desert, where drinking water is available. Absent from tropical regions
HEAD AND BODY LENGTH: 57–74 cm
TAIL LENGTH: 36–45 cm
DISTRIBUTION: more than 1 million km^2
ABUNDANCE: abundant
STATUS: secure

(LF Schick)

Order LAGOMORPHA
(lah'-goh-mor'-fah: "hare-shaped")

Like rodents, the members of this group have continuously growing upper and lower incisors that are self-sharpening as they bite against each other, but lagomorphs have two pairs of upper incisors (one pair in rodents). The ears are large, the eyes are set high on the head, and the slit-like nostrils can be closed. The tail is extremely short. Like ringtailed possums, lagomorphs manage digestion of plant fibre by producing a soft faeces which is eaten and passes through the gut a second time, the final faeces being a compact fibrous pellet.

The order comprises two families: the Ochotonidae, with 14 species of pikas; and the Leporidae, with 44 species of rabbits and hares. They are distributed naturally over all continents except Australia.

Family LEPORIDAE
(lep-or'-id-ee: "*Lepus*-family", after a genus of hares)

This family includes the hares and rabbits. Both have long ears and long hind legs. The basic difference between the two groups is that rabbits tend to be specialised for burrowing, whereas hares escape their predators by running (often in a confusing zigzag).

Genus Lepus
(lep'-us: "hare")

In general, and particularly in respect of the species occurring in Australia, hares have longer ears and longer legs than rabbits. The Y-shaped groove from the centre of the upper lip to the nostrils ("hare-lip") is more pronounced.

Brown Hare
Lepus capensis (kayp-en'-sis: "Cape [-of-Good-Hope] hare")

(*LF Schick*)

HABITAT: cool to warm temperate farmland and woodland
HEAD AND BODY LENGTH: 50–60 cm
TAIL LENGTH: 7–9 cm
DISTRIBUTION: more than 1 million km²
ABUNDANCE: sparse to common
STATUS: secure

The natural range of the Brown Hare appears to extend from much of China to Spain and Africa. If, as some authorities claim, it is not distinct from the European Hare (*L. europaeus*) its range also extends to western and northern Europe. The founders of the Australian population were introduced from Europe as game animals in the 1860s and spread from Victoria through much of south-eastern Australia and Tasmania. It is solitary, resting during the day in a scrape under foliage or grass and emerging in the evening and at night to graze, or sometimes to browse.

Sexual maturity is reached in the first year. Mating can occur at any time but is mainly from June to September. After a gestation of six weeks, two to five young are born. Up to six litters may be reared in a year.

Genus Oryctolagus

(oh-rik'-toh-lah'-gus: "burrowing-hare")

The characteristics of this genus are those of the single species.

European Rabbit

Oryctolagus cuniculus (kue-nik'-ue-lus: "rabbit burrowing-hare")

The original home of the European Rabbit seems to have been Spain and North Africa. The species was domesticated between the fifth and 10th centuries AD but became feral over much of Europe. The progenitors of the Australian population were feral animals from Spain, deliberately set free around the 1860s as game. By the 1920s it had occupied about three-quarters of the continent, absent only from the tropics. The Rabbit is a grazer but will browse when grasses are scarce. If its food is sufficiently succulent, it does not need to drink. Rabbits are gregarious, spending the day in multiple, common burrows (warrens) and feeding at night.

Sexual maturity is attained at three to four months. After a gestation of 30 days, a litter of up to five is born. As many as five litters may be raised in a year.

(PW Menkhorst)

HABITAT: Almost any temperate to subtropical environment except rainforest
HEAD AND BODY LENGTH: 36–43 cm
TAIL LENGTH: *c* 5 cm
DISTRIBUTION: more than 1 million km²
ABUNDANCE: abundant
STATUS: secure

Genus Mus

(moos: "mouse")

It is impossible to differentiate this genus from the many other similar genera without recourse to fine anatomical details. It is "mouse-like".

House Mouse

Mus musculus (moos'-kue-lus: "mouse mouse")

It seems that this species originated in central Asia but it has spread over most of the world, largely in association with human habitations. It probably arrived in Australia with the First Fleet, or even earlier, and has spread far from cities and towns: it may now be the most widespread and numerous of all the mammals on the continent. It is particularly prevalent where cereals are farmed in the areas of regrowth after bushfires. The diet includes food scraps, stored foods, seeds and insects.

Sexual maturity is attained at about two months. There is no particular breeding season but reproductive activity is reduced in times of food shortage. After a gestation of less than three weeks, a litter of four to eight young is born. In favourable times, a female can rear 11 litters in a year. Particularly in inland areas, the House Mouse may vary dramatically in numbers. When food is abundant, populations may increase to plague proportions. "Plagues" usually represent the last stages of a population explosion, just prior to a dramatic decrease in numbers.

(R Whitford)

HABITAT: cool temperate to tropical areas of almost any nature
HEAD AND BODY LENGTH: 6.0–9.5 cm
TAIL LENGTH: 7.6–9.5 cm
DISTRIBUTION: more than 1 million km²
ABUNDANCE: abundant
STATUS: secure

Brown Rat

Rattus norvegicus (nor-veg'-ik-us: "Norway rat")

Originally from the Caspian region, the Brown Rat has been transported around the world by humans, probably reaching Australia with the first European settlers. It is readily distinguished from the Black Rat by its relatively shorter tail. In Australia, it is largely limited to the vicinity of human habitations in relatively well-watered areas. It feeds at night, mainly on food scraps and stored foods, but also eats seeds, insects, bird eggs and young birds. Social organisation is loose but is based upon several females and their young under the dominance of a male.

Sexual maturity is reached at three to four months. A litter of seven to 10 young is born after a gestation of about three weeks and as many as six litters may be reared in a year.

(CA Henley)

HABITAT: cool temperate to sub-tropical human habitations in well-watered regions, particularly ports; river banks
HEAD AND BODY LENGTH: 18–26 cm
TAIL LENGTH: 15–20 cm
DISTRIBUTION: 100,000–300,000 km²
ABUNDANCE: common
STATUS: secure

Black Rat

Rattus rattus (rat'-us: "rat rat")

The original home of this species appears to have been in the Middle East but it was accidentally introduced to Europe in the 13th century by returning Crusaders. It is likely that it was introduced to Australia (also accidentally) by the first European settlers. It lives in cities and towns around most of the well-watered coastal regions of Australia. It sleeps by day in a nest in a secluded place, emerging at night to feed on food scraps, carrion and a wide range of plant material, including fruits.

Sexual maturity is reached at three to four months. A litter of five to 10 young is born after a gestation of three weeks. As many as six litters may be reared in a year.

HABITAT: cool temperate to tropical well-watered regions with
HEAD AND BODY LENGTH: 16–21 cm
TAIL LENGTH: 18–25 cm
DISTRIBUTION: 300,000–1 million km²
ABUNDANCE: common
STATUS: secure

(B & B Wells)

CLASS MAMMALIA

SUBCLASS PROTOTHERIA
ORDER MONOTREMATA
> Family Ornithorhynchidae
>> Genus *Ornithorhynchus* Platypus
> Family Tachyglossidae
>> Genus *Tachyglossus* Short-beaked Echidna

SUBCLASS MARSUPIALIA
COHORT AUSTRALIDELPHIA
ORDER DASYUROMORPHIA
> Family Dasyuridae
>> Subfamily Dasyurinae
>>> Genus *Dasycercus* Mulgara
>>> Genus *Dasykaluta* Kaluta
>>> Genus *Dasyuroides* Kowari
>>> Genus *Dasyurus* quolls
>>> Genus *Parantechinus* dibblers
>>> Genus *Pseudantechinus* pseudantechinuses
>>> Genus *Sarcophilus* Tasmanian Devil
>> Subfamily Phascogalinae
>>> Genus *Antechinus* antechinuses
>>> Genus *Phascogale* phascogales
>> Subfamily Planigalinae
>>> Genus *Ningaui* ningauis
>>> Genus *Planigale* planigales
>> Subfamily Sminthopsinae
>>> Genus *Antechinomys* Kultarr
>>> Genus *Sminthopsis* dunnarts
> Family Myrmecobiidae
>> Genus *Myrmecobius* Numbat
> Family Thylacinidae
>> Genus *Thylacinus* Thylacine

ORDER PERAMELOMORPHIA
> Superfamily Perameloidea
> Family Peramelidae
>> Genus *Chaeropus* Pig-footed Bandicoot
>> Genus *Isoodon* short-nosed bandicoots
>> Genus *Macrotis* bilbies
>> Genus *Perameles* long-nosed bandicoots
> Family Peroryctidae
>> Genus *Echymipera* spiny bandicoots

ORDER NOTORYCTEMORPHIA
> Family Notoryctidae
>> Genus *Notoryctes* Marsupial Mole

ORDER DIPROTODONTIA
SUBORDER VOMBATIFORMES
INFRAORDER VOMBATOMORPHIA
> Family Vombatidae
>> Genus *Lasiorhinus* hairy-nosed wombats
>> Genus *Vombatus* Common Wombat
INFRAORDER PHASCOLARCTOMORPHIA
> Family Phascolarctidae
>> Genus *Phascolarctos* Koala
SUBORDER PHALANGERIDA
> Superfamily Phalangeroidea
> Family Phalangeridae
>> Genus *Phalanger* cuscuses
>> Genus *Trichosurus* brushtail possums
>> Genus *Wyulda* Scaly-tailed Possum
> Superfamily Burramyoidea
> Family Burramyidae
>> Genus *Burramys* Mountain Pygmy-possum
>> Genus *Cercartetus* pygmy-possums
> Superfamily Petauroidea
> Family Petauridae
>> Genus *Dactylopsila* striped possums
>> Genus *Gymnobelideus* Leadbeater's Possum
>> Genus *Petaurus* gliders
> Family Pseudocheiridae
>> Genus *Hemibelideus* Lemuroid Ringtail
>> Genus *Petauroides* Greater Glider
>> Genus *Pseudocheirus* ringtail possums
> Superfamily Tarsipedoidea
> Family Acrobatidae
>> Genus *Acrobates* Feathertail Glider
> Family Tarsipedidae
>> Genus *Tarsipes* Honey-possum
> Superfamily Macropodoidea
> Family Potoroidae
>> Subfamily Hypsiprymnodontinae
>>> Genus *Hyspiprymnodon* Musky Rat-kangaroo
>> Subfamily Potoroinae
>>> Genus *Aepyprymnus* Rufous Bettong

Genus *Bettongia* bettongs
Genus *Caloprymnus* Desert Rat-kangaroo
Genus *Potorous* potoroos
Family Macropodidae
Subfamily Macropodinae
Genus *Dendrolagus* tree-kangaroos
Genus *Lagorchestes* hare-wallabies
Genus *Macropus* wallabies, wallaroos
kangaroos
Genus *Onychogalea* nailtail wallabies
Genus *Peradorcas* Nabarlek
Genus *Petrogale* rock-wallabies
Genus *Thylogale* pademelons
Genus *Wallabia* Swamp Wallaby
Subfamily Sthenurinae
Genus *Lagostrophus* Banded Hare-wallaby

SUBCLASS EUTHERIA
ORDER CHIROPTERA
SUBORDER MEGACHIROPTERA
Family Pteropodidae
Subfamily Macroglossinae
Genus *Macroglossus* blossom-bats
Genus *Syconycteris* blossom-bats
Subfamily Nyctimeninae
Genus *Nyctimene* tube-nosed bats
Subfamily Pteropodinae
Genus *Dobsonia* bare-backed fruit-bats
Genus *Pteropus* flying foxes
SUBORDER MICROCHIROPTERA
Family Emballonuridae
Genus *Saccolaimus* sheathtail-bats
Genus *Taphozous* sheathtail-bats
Family Megadermatidae
Genus *Macroderma* Ghost Bat
Family Molossidae
Genus *Chaerophon* mastiff-bats
Genus *Mormopterus* mastiff-bats
Genus *Nyctinomus* mastiff-bats
Family Rhinolophidae
Genus *Hipposideros* horseshoe-bats
Genus *Rhinolophus* horseshoe-bats
Genus *Rhinonicteris* horseshoe-bats
Family Vespertilionidae
Subfamily Kerivoulinae
Genus *Kerivoula* groove-toothed bats

Subfamily Miniopterinae
Genus *Miniopterus* bent-wing bats
Subfamily Murininae
Genus *Murina* tube-nosed insectivorous bats
Subfamily Nyctophilinae
Genus *Nyctophilus* long-eared bats
Subfamily Vespertilioninae
Genus *Chalinolobus* pied and wattled bats,
eptesicuses, pipistrelles, myotises,
broad-nosed bats

ORDER RODENTIA
SUBORDER MYOMORPHA
Family Muridae
Subfamily Hydromyinae
Tribe Conilurini
Genus *Conilurus* rabbit-rats
Genus *Leggadina* native mice
Genus *Leporillus* stick-nest rats
Genus *Mastacomys* Broad-toothed Rat
Genus *Mesembriomys* tree-rats
Genus *Notomys* hopping-mice
Genus *Pseudomys* native mice
Genus *Zyzomys* rock-rats
Tribe Hydromyini
Genus *Hydromys* water-rats
Genus *Xeromys* False Water-rat
Tribe Uromyini
Genus *Melomys* melomyses
Genus *Uromys* giant rats
Genus *Pogonomys* prehensile-tailed rats
Subfamily Murinae
Genus *Rattus* 'true' rats
Genus *Mus* 'true' mice

ORDER CARNIVORA
Family Canidae
Genus *Canis* dogs
Family Otariidae
Genus *Arctocephalus* fur-seals
Genus *Neophoca* Australian Sea-lion

ORDER SIRENIA
Family Dugongidae
Genus *Dugong* Dugong

Introduced Species

ORDER PERISSODACTYLA
    Family Equidae
        Genus *Equus* Horse, Donkey

ORDER ARTIODACTYLA
    Family Bovidae
        Genus *Bos* Banteng
        Genus *Bubalus* Water Buffalo
    Family Capridae
        Genus *Capra* Goat
    Family Camelidae
        Genus *Camelus* Camel
    Family Cervidae
        Genus *Axis* deer
        Genus *Cervus* deer
        Genus *Dama* deer
    Family Suidae
        Genus *Sus* Pig

ORDER RODENTIA

ORDER CARNIVORA
    Family Canidae
        Genus *Vulpes* Fox

ORDER LAGOMORPHA
    Family Leporidae
        Genus *Lepus* Hare
        Genus *Oryctolagus* Rabbit

GLOSSARY

Amphibious. Able to live on land and in water.

Allopatric. Living in different areas. Usually applied to congeneric species. Contrast with **sympatric**.

Arboreal. Living in trees.

Arthropods. A wide variety of animals with external skeletons and jointed bodies and limbs: e.g. crustaceans, millipedes, centipedes, insects, spiders.

Blastocyst. An early stage in the development of a mammal when the embryo is a tiny, undifferentiated ball of cells.

Brigalow. An Australian wattle, *Acacia harpophylla*. Also areas where this species is dominant in the vegetation.

Caecum. A blind branch of the intestine, often large in herbivorous mammals.

Calcrete. Hard soil composed of gravel, sand or desert debris, cemented by porous calcium carbonate.

Canopy. The upper foliage, usually dense, of a tree or forest.

Carnivorous. Feeding on animals (usually vertebrates).

Caudal. Pertaining to the tail.

Chenopod. Any plant of the family Chenopodiaceae: e.g. saltbush and bluebush.

Cline. A gradual change in the characteristics of a species across its range—usually associated with an environmental gradient: north–south, wet–dry, lowland–mountain, etc. Contrast with **subspecies**.

Cloaca. An aperture in female vertebrates (except eutherian mammals) through which faeces, urine and eggs pass out of the body. In male vertebrates that have a cloaca, this serves for the exit of faeces and urine, but sperms may be ejected through a penis. The cloacal aperture is often referred to, inaccurately, as the anus.

Congeners. Members of the same genus. Species in the same genus are said to be congeneric.

Conspecific. Belonging to the same species.

Convergent evolution. The evolution of similar body form (or other features) in animals of widely different ancestry: e.g. sharks and dolphins, bats and birds.

Cryptic. Hidden, inconspicuous, skilled at concealment.

Diapause. A temporary halt in embryonic or larval development.

Diastema. A gap in a series of teeth, usually between the incisors at the front of the mouth and the grinding teeth at the rear.

Distribution. The maximum extent of the area in which a species is known to occur, also referred to as range. See also **home range, territory**.

Dorsal. Pertaining to the back or upper surface of an animal. Opposite of **ventral**.

Emergent tree. One which rises well above the surrounding forest canopy.

Endemic. Native to a designated area: e.g. the Platypus is endemic to Australia. See also **exotic** and **extralimital**.

Fossorial. Burrowing. Spending all or part of the time below ground.

Exotic. Foreign, coming from outside a designated area: e.g. Cane Toads and Camels are exotic in respect of Australia.

Extralimital. Living inside and outside of a designated area: e.g. the Spotted Cuscus, which lives in Australia and New Guinea, is extralimital in respect of Australia.

Exudate. An exuded substance. In this context, usually refers to edible sap and gum produced by plants and the nutritious excreta of certain sap-sucking insects.

Forb. Any small, non-woody, ground plant which is not a grass. Plants designated as forbs usually have broad leaves.

Friable. Easily crumbled: descriptive of soils.

Frugivorous. Feeding on fruits.

Gibber desert. Desert with a surface composed largely of smoothly rounded stones (gibbers).

Gilgai. A soil formation with an undulating surface, often with mounds and depressions caused by seasonal swelling and contraction of clay. Also known as "crab-hole country".

Habitat. The area that provides the physical and biological requirements of a species.

Herbivorous. Feeding on plants.

Hibernation. A prolonged condition of deep sleep and diminished temperature control during winter. See also **torpor**.

Home range. The area habitually traversed by an individual animal. It may be exclusive or overlap with the home ranges of other members of the species.

Hummock grassland. Areas where grasses of the genus *Triodia* and *Plectrachne* (both often erroneously called "spinifex") are dominant. Soil builds up around each clump of grass, forming a hummock, the ground between the hummocks usually being bare. See also **spinifex**.

Hybrid. The offspring of individuals from two distinct populations. As a general rule, hybrids between related subspecies are viable and fertile; hybrids between different species are rare in nature and seldom fertile.

Insectivorous. Feeding on insects and/or other terrestrial arthropods.

Invertebrate. Any animal that is not a vertebrate: worms, molluscs, arthropods, etc.

Labial. Referring to the lips.

Lateral. Referring to the sides.

Lateritic. Pertaining to laterite, a reddish soil derived from the breakdown of rocks rich in iron. Laterite has a hard surface crust.

Lignum. *Muehlenbeckia cunninghami*, a sparse-leafed shrub which grows into a tangled mound of thin canes, often forming dense thickets.

Lingual. Pertaining to the tongue.

Littoral. Pertaining to the edge of a sea or lake.

Mallee. Eucalypt trees with multiple stems arising from a single root stock. Also country in which these trees are dominant in the vegetation.

Melanic, melanotic. Black or darkly pigmented. Usually descriptive of a morph.

Mesic. Pertaining to an environment intermediate between arid and moist.

Migratory. Moving more or less regularly from one area to another, usually in response to seasonal change. Compare with **sedentary** and **nomadic**.

Monophyletic. Sharing a common ancestor in the group of organisms under consideration. Contrast with **polyphyletic**.

Monsoon rainforest. Low rainforest occurring in small patches across northern Australia from Cape York to the Kimberleys. It is lower and less complex than typical tropical rainforest.

Morph. One of two or more distinct forms (usually differing in coloration) which occur within a freely interbreeding population. See also **polymorphic**.

Nectarivorous. Feeding on nectar (sometimes also pollen).

Nomadic. Wandering from place to place, usually from one source of food to another. Contrast with **sedentary** and **migratory**.

Omnivorous. Feeding on animals and plants.

Opportunistic. In reference to feeding behaviour, eating whatever is available within a wide range of foods. In reference to reproduction, breeding when environmental conditions are favourable, rather than at a particular time of year.

Oviparous. Egg-laying. Compare with **ovoviviparous** and **viviparous**.

Ovoviviparous. Producing eggs that are retained in the body until they hatch. Compare with **oviparous**.

Patagium. A gliding membrane, usually extending between the body and the legs on each side.

Pelagic. Living at or near the surface of the sea.

Peneplain. An area reduced by erosion almost to a level plain.

Pinna. That part of the ear that projects outwards from the head. Usually referred to simply as the "ear".

Polymorphic. The condition in which a species or population consists of two or more morphs which are distinct in coloration (or other characters). A population with two morphs is said to be dimorphic.

Polyphyletic. Not sharing a common ancestor within the group under consideration. Contrast with **monophyletic**.

Posterior. Pertaining to the rear end of an animal. Opposite of anterior.

Prehensile. Able to grip. Refers to a limb or tail.

Proximal. At, or towards the region of attachment of, an appendage such as a limb or tail. Opposite of distal.

Refugial. Referring to a limited area in which a sometimes widespread species survives during unfavourable conditions such as drought. Such an area is called a refugium.

Relic, relictual. Refers to the isolated remnant population(s) of a species that was once more widely distributed.

Riparian. Pertaining to the land on the sides of a river: riverside.

Rostral. On, or towards, the snout.

Sclerophyll. A general term for eucalypt forest or woodland. Wet sclerophyll forest is tall and has a dense canopy. Dry sclerophyll forest varies considerably in height and usually has a discontinuous canopy.

Sedentary. Remaining in much the same area throughout the year. Contrast with **migratory** and **nomadic**.

Senescent. Becoming senile, approaching the end of life.

Spatulate. Flattened at the tip; usually in reference to fingers or toes.

Spinifex. Strictly, a genus of sea-coast grasses with long, creeping stems. Commonly, and inaccurately, used as a general term for spiky hummock grasses (*Triodia, Plectrachne*) of the arid inland. See **hummock grassland**.

Subspecies. One of two or more populations of a species that are recognisably different from each other and usually differ in distribution. Subspecies frequently hybridise where their distributions meet or overlap.

Substrate. The surface—soil, rock, leaf-litter, etc.—on which an animal lives.

Swale. A more or less flat area between dunes.

Sympatric. Living in the same area. Usually applied to congeneric species. Contrast with **allopatric**.

Taxon. The scientific name of any classificatory group of organisms: e.g., phylum, class, order, family, genus, species.

Terrestrial. Pertaining to the land. Living on, or mainly on, the ground. Contrast with **arboreal, amphibious, fossorial**.

Territory. An area occupied by one or more individuals and defended against other members of the species.

Torpor. A state of dormancy and diminished temperature regulation over a period of hours or days, in response to cold or food shortage. See also **hibernation**.

Tussock grassland. Area where Mitchell Grass (*Astrebla*) or bluegrass (*Dicanthium*) are the dominant plants. See also **spinifex** and **hummock grassland**.

Ventral. Pertaining to the belly or under-surface. Opposite to **dorsal**.

Vine thicket. Rainforest with many hanging vines or lianas.

Viviparous. Giving birth to live young.

INDEX OF COMMON NAMES

INDEX OF SCIENTIFIC NAMES

Perameles
bougainville 37
eremiana 38
gunnii 38
nasuta 39

Petauroides
volans 57

Petaurus
australis 54
breviceps 55
norfolcensis 55

Petrogale
brachyotis 81
burbidgei 81
godmani 82
inornata 82
lateralis 83
penicillata 83
persephone 84
rothschildi 84
xanthopus 85

Phalanger
maculatus 44
orientalis 45

Phascogale
calura 18
tapoatafa 19

Phascolarctos
cinereus 41

Pipistrellus
adamsi 123
westralis 124

Planigale
gilesi 21
ingrami 21
maculata 22
tenuirostris 22

Pogonomys
mollipilosus 152

Potorous
longipes 67
platyops 67
tridactylus 68

Pseudantechinus
macdonnellensis 12
ningbing 12
woolleyae 13

Pseudocheirus
archeri 58
dahli 59
herbertensis 59
peregrinus 60

Pseudomys
albocinereus 136
apodemoides 136
australis 137
bolami 137
chapmani 137
delicatulus 138
desertor 138
fumeus 139
gouldii 139
gracilicaudatus 140
hermannsburgensis 140
higginsi 141
johnsoni 141
laborifex 142
nanus 142
novaehollandiae 143
occidentalis 143
oralis 144
pilligaensis 144
praeconis 145
shortridgei 145

Pteropus
alecto 94
conspicillatus 95
poliocephalus 95
scapulatus 96

Rattus
colletti 153
fuscipes 153
leucopus 154
lutreolus 154
norvegicus 175
rattus 175
sordidus 155
tunneyi 155
villosissimus 156

Rhinolophus
megaphyllus 108
philippinensis 108

Rhinonicteris
aurantius 109

Saccolaimus
flaviventris 97
mixtus 98
saccolaimus 98

Sarcophilus
harrisii 13

Scoteanax
rueppellii 124

Scotorepens
balstoni 125
greyii 125
orion 126
sanborni 126

Setonix
brachyurus 85

Sminthopsis
aitkeni 23
archeri 24
butleri 24
crassicaudata 25
dolichura 25
douglasi 26
gilberti 26
granulipes 26
griseoventer 27
hirtipes 27
leucopus 28
longicaudata 28
macroura 29
murina 29
ooldea 30
psammophila 30
virginiae 31
youngsoni 31

Sus
scrofa 171

Syconycteris
australis 92

Tachyglossus
aculeatus 5

Taphozous
australis 99
georgianus 99
hilli 100
kapalgensis 100

Tarsipes
rostratus 62

Thylacinus
cynocephalus 33

Thylogale
billardierii 86
stigmatica 86
thetis 87

Trichosurus
arnhemensis 46
caninus 46
vulpecula 47

Uromys
caudimaculatus 151

Vombatus
ursinus 43

Vulpes
vulpes 172

Wallabia
bicolor 87

Wyulda
squamicaudata 48

Xeromys
myoides 148

Zyzomys
argurus 146
pedunculatus 146
woodwardi 147